Understanding Inconsistent Science

Understanding Inconsistent Science

Peter Vickers

UNIVERSITY PRESS

Great Clarendon Street, Oxford, OX2 6DP,
United Kingdom

Oxford University Press is a department of the University of Oxford.
It furthers the University's objective of excellence in research, scholarship,
and education by publishing worldwide. Oxford is a registered trade mark of
Oxford University Press in the UK and in certain other countries

© Peter Vickers 2013

The moral rights of the author have been asserted

First Edition published in 2013

Impression: 1

All rights reserved. No part of this publication may be reproduced, stored in
a retrieval system, or transmitted, in any form or by any means, without the
prior permission in writing of Oxford University Press, or as expressly permitted
by law, by licence or under terms agreed with the appropriate reprographics
rights organization. Enquiries concerning reproduction outside the scope of the
above should be sent to the Rights Department, Oxford University Press, at the
address above

You must not circulate this work in any other form
and you must impose this same condition on any acquirer

British Library Cataloguing in Publication Data
Data available

ISBN 978-0-19-969202-6

Printed in Great Britain by
CPI Group (UK) Ltd, Croydon, CR0 4YY

Preface

An inconsistent scientific theory is a theory which contradicts itself. Speaking in the common way (for now), it is important for scientists to notice when a given theory is inconsistent, and when it is not. For example, we know that an inconsistent theory is a false theory without having to go to the trouble of submitting it to experimental test. Thus the practice of scientists can be dramatically affected when a theory is found to be inconsistent; in particular, experiments designed to test its truth-candidacy can be halted. More resources might be put into altering the theory, or devising a new theory altogether.

But over the past sixty years there have been an increasing number of claims that scientists sometimes do work with internally inconsistent theories, even knowingly, and that such science can be extremely successful. This has led to increased interest in inconsistency in science, as philosophers attempt to answer various questions. One can distinguish two major debates: (i) the logic-driven versus content-driven debate, and (ii) the debate as to how inconsistent theories should be represented. In the former debate one asks questions about how scientists both *do* and *should* reason with inconsistent scientific theories: do/should they adjust their logic to avoid deriving absurd results, or do/should they restrict their inferences by looking to the material content of the theory? The other debate starts by noting that an inconsistent theory cannot possibly be represented as a set of propositions and its consequences (this will be the set of all propositions!), and considers different ways in which such a theory might be represented.

The present investigation starts by noting that both of these debates put scientific theories at the centre of the picture: the issue is how we reason with, or represent, inconsistent scientific theories. Thus how we think about scientific theories in the first place is going to affect how the debates are framed. One might even claim that both debates are wrong-headed from the start, because scientific theories aren't even the *kind* of thing which can be consistent or inconsistent (as some advocates of the semantic approach to theories have suggested). At the very least, different philosophers engaged in these debates will disagree with each other in one way or another about 'what theories are' (including the content that should be ascribed to specific theories). These disagreements will then lead to disagreements about inconsistency in science.

My initial response to this situation was to first take a stance on what theories are, or at least how they should be conceived, and then assess alleged cases of inconsistency in science accordingly. But I was conscious that I could find myself in a situation (as others have) where many of my conclusions about inconsistency would fall down if my own 'theory of theories' did. At this point I made a decision to place very little weight on the concept of a scientific theory. I would assert that theories do have a minimal 'essence' or 'core' (following Gould 2002 and Morrison 2007), but I would say as little as possible about additional content. But it turns out that this 'minimal' approach also has very significant drawbacks.

Nevertheless, thinking about the 'minimal' approach led me to the idea that I might be able to move forward without putting *any* weight on the concept *scientific theory*, making no claims about the nature of scientific theories whatsoever. Roughly speaking the approach is as follows: one simply presents what *is* inconsistent in a given case, and then assesses what (if anything) that inconsistency means for science and philosophy of science by examining the nature and historical relevance of the constituents in play. For the particular example of asking questions about *inconsistency* in science the constituents in play will be *propositions* (broadly construed). The important questions are then 'What exactly do the propositions say about the world?', 'What is the point of grouping them together?', 'How did/do scientists commit to them?', 'How did/do scientists use them?' and so on. It is by asking questions such as these that the character of, and correct response to, a given inconsistency is unveiled. An answer to the question 'But is that really *the theory* you're talking about?' just isn't needed to learn the lessons about inconsistency in science that really matter.

So, in the case studies to follow, I simply stay silent on questions about the nature of theories. It may or may not be that theories are made up of 'statements' or 'models'; it may or may not be that a given theory has a definitive content, or even a definitive 'core'. I argue—and show through specific case studies—that all of the really important questions about inconsistency in science can be asked and answered without taking a stance on such issues. The consequence is that many disagreements in the current literature (e.g. about whether some theory is *really* inconsistent or not) cease to be problematic, or even interesting. Of course, on the face of it, it may seem an impossible job to assess claims and disagreements in the literature about the consistency *of scientific theories* without taking any stance myself on the nature of scientific theories. Nevertheless, in what follows I hope to convince the reader not only that this can be done, but that the benefits are considerable. The proof will be in the pudding.

With the method vis-à-vis the concept *scientific theory* presented in Chapter 2, the rest of the book is devoted to a reappraisal of the 'internally inconsistent scientific theories' regularly referred to in the literature. The monograph devotes full chapters to Bohr's theory of the atom, classical electrodynamics, Newtonian cosmology, and the early calculus. It then turns more briefly, in Chapter 7, to Aristotle's theory of motion, Olbers' paradox, classical theories of the electron, and Kirchhoff's theory of the diffraction of light. Up until now these have been grouped together as 'inconsistent scientific theories', a common assumption being that since they are all the same type of thing (all *theories*), the inconsistency will manifest itself in each of them in basically the same way, so that one can say something substantial and general about inconsistency in science. What the method of Chapter 2 shows up is that, owing to the idiosyncrasies of these cases and the individual characters of the (sets of) propositions in play, the inconsistency means something significantly different in each case. Thinking in terms of 'inconsistent theories' assumes from the start a uniformity to science that doesn't do the historical and scientific facts justice. Accordingly, no general 'theory of reasoning with inconsistency' or 'theory of inconsistency resolution' is to be expected, although there are interesting overlaps and interrelations between the different cases (collected together and discussed in Chapter 8).

Ultimately the monograph has three goals: (i) to stand as a critical examination of the foundations of studies into inconsistency in science, (ii) to show how philosophy of science can be done without putting analytical weight on the complex and troublesome concept *scientific theory*, and (iii) to stand as a test case for examining the current state of play vis-à-vis the relationship between philosophy of science, history of science, and science itself. Thus although many of the conclusions pertain only to the narrow topic of inconsistency in science, in other ways the goal is to examine certain aspects of the method of philosophy of science in general. In this way the consequences potentially reach right across the discipline.

This book has its origins in my PhD thesis, and thus has been written gradually over a period of several years. The thesis itself was written during my postgraduate years at the University of Leeds, UK, whereas much of the development into a monograph took place during my year as a Postdoctoral Fellow at the Center for Philosophy of Science at the University of Pittsburgh, USA. I am indebted to both institutions for providing exceptionally stimulating environments in which to produce the work contained herein. I am also exceedingly grateful for an AHRC grant for the first three years of the PhD, and a Royal Institute of Philosophy Jacobsen Fellowship for the final year of the PhD, without either of which this work simply wouldn't exist. In addition, very special thanks go to Steven French, John Norton, Mathias Frisch, Gordon Belot, Chris Timpson, Juha Saatsi, Otávio Bueno, Joseph Melia, Edouard Machery, Mark Wilson, Matthew Kieran, Robin Hendry, Darrell Rowbottom, Alan Chalmers, and two anonymous referees, as well as numerous other friends and associates in the field who provided inspiration in one way or another.

Last, but absolutely not least, I must mention my closest family, who have supported me unconditionally. My parents John and Teresa Vickers have worked so hard for so long to allow me the opportunities they never had. Despite coming from dramatically different walks of life (even in one generation!) they have never questioned my decision to pursue an academic career in an abstract discipline. They provided the foundation for the writing of this book, but it is my wife Laura who walked every step of it with me. She has shown astonishing patience, and made dramatic sacrifices (which ought never to have been asked of her). This book belongs to her more than anyone. She deserves for it to sell a million copies.

Contents

1 Introduction   1
   1.1 Who Cares about Inconsistency Anyway?   1
   1.2 Inconsistency in Science: Contemporary Debates   6
      1.2.1 Logic-driven versus content-driven   6
      1.2.2 Reconstructing/representing inconsistencies   10
   1.3 Going Deeper   12

2 Concepts and Method   17
   2.1 Introduction   17
   2.2 What is Inconsistency?   17
   2.3 Theories of Theories: Syntactic and Semantic   19
   2.4 The Substrate of Science   22
   2.5 Concepts and Contents of Theories   25
   2.6 A New Method: Theory Eliminativism   28
   2.7 Back to Inconsistency   32
   2.8 Overview   35

3 Bohr's Theory of the Atom   39
   3.1 Introduction   39
   3.2 Three Inconsistency Claims   42
      3.2.1 Discrete energy states   43
      3.2.2 Quantum transitions   46
      3.2.3 Non-emission of radiation   48
         3.2.3.1 Doxastic commitment   49
         3.2.3.2 Inferential confidence   52
   3.3 Inconsistencies in the Later Theory   58
      3.3.1 Spectral line intensities   59
      3.3.2 The adiabatic principle   64
   3.4 Conclusion   71

4 Classical Electrodynamics   76
   4.1 Introduction   76
   4.2 Features of the Theory   76

	4.3 Frisch's Inconsistency Claim	82
	4.4 Defending Frisch	84
	4.4.1 Muller	85
	4.4.2 Belot	86
	4.5 The Significance of the Frisch Inconsistency	99
	4.6 Conclusion	105
5	Newtonian Cosmology	110
	5.1 Introduction	110
	5.2 The Concept *Newtonian Cosmology*	111
	5.3 How was Newtonian Cosmology Inconsistent?	113
	5.3.1 A contradiction of forces	114
	5.3.1.1 ... using Newton's law of gravitation?	114
	5.3.1.2 ... using Poisson's equation?	118
	5.3.1.3 ... reasoning from symmetry	120
	5.3.2 An Indeterminacy Contradiction	124
	5.3.2.1 ... using Newton's law of gravitation	125
	5.3.2.2 ... from summing the potential φ	128
	5.3.2.3 ... using Poisson's equation	130
	5.4 Why weren't the Inconsistencies Noticed?	132
	5.4.1 Because the right question wasn't asked	133
	5.4.2 Because of confusion about non-convergent series	137
	5.5 Conclusion	142
6	The Early Calculus	146
	6.1 Introduction	146
	6.2 Berkeley's Criticism	147
	6.3 The Algorithmic Level	150
	6.4 The Level of Justification	156
	6.4.1 Newton	158
	6.4.2 Leibniz	165
	6.4.3 The English	169
	6.4.4 The French	174
	6.5 'As if' Inconsistency	182
	6.6 Conclusion	188
7	Further Examples	192
	7.1 Introduction	192
	7.2 Aristotle's Theory of Motion	192

	7.3 Olbers' Paradox	196
	7.4 Classical Electrons	199
	7.4.1 Reconstruction of the inconsistency	201
	7.4.2 What lessons?	203
	7.5 Kirchhoff's Theory of Diffraction	208
	7.5.1 Reconstruction of the inconsistency	209
	7.5.2 What lessons?	212
	7.6 Conclusion	216
8	Conclusion	218
	8.1 Introduction	218
	8.2 Lessons from the Case Studies	219
	8.2.1 Asking questions	219
	8.2.2 Inconsistency versus implausibility	226
	8.2.3 The role of 'as if' reasoning	230
	8.2.4 Getting from inconsistency to contradiction	233
	8.2.5 The logic-driven/content-driven debate	238
	8.3 Theory Eliminativism and the Method of Philosophy of Science	242
	8.4 Concluding Thoughts	252

Bibliography   254
Index   269

1
Introduction

1.1 Who Cares about Inconsistency Anyway?

Inconsistency is a powerful thing. When we find something to be inconsistent it means that it is self-contradictory, and thus tells us unequivocally that there is something wrong. If two criminals get confused about a previously agreed alibi, with one saying that the two of them spent the entire evening in question together at 36 Farrar Street, and the other saying that the two of them spent the entire evening together at 23 Milson Grove, then we know immediately that one or both of them is speaking falsely. Their stories are inconsistent with each other, so they cannot possibly both be true. This is obvious and uncontroversial, so that when they realize what they've done there is no chance whatsoever that either person will maintain that *both* of them are telling the truth (unless they wish to plead insanity).

On the other hand, if their alibis are *consistent*—for example because they are exactly the same—then we are none the wiser as to whether they are speaking truth or falsehood. All we can do is go out into the world and investigate their claims. And even then, whatever we find, we will never know *for sure* that what they say is true or that what they say is false. We may consider it beyond all reasonable doubt that they are lying, but we do not know. We may *say* we 'know' they are speaking falsely, for example because we get DNA evidence, CCTV evidence, and multiple witness reports contrary to the alibi. But, strictly speaking, it remains possible that they are telling the truth and all of this 'evidence' is part of a massive conspiracy to frame them. There is no similar story with inconsistency: with inconsistency we know *for a fact* that at least one of them is speaking falsely. Thus inconsistency can be much more powerful, and much more useful, than consistency.

Philosophy makes a special effort to discover and attempt to resolve inconsistencies in our belief sets. In a recent 'philosophical quiz book', *Do You Think What You Think You Think?* (Baggini and Stangroom 2006), the authors introduce philosophy in the first chapter by attempting to make readers contradict themselves. Anyone who has done a lot of philosophy will no doubt have been moved at one time or another by finding a tension in their own belief set. Indeed, for many finding such a tension is the inspiration to study philosophy in the first place. My favourite example is to encourage students to agree that, if they were 'beamed down' to another planet, Star-Trek-style, then they would be the same person after the trip. After all, they would be *exactly the same* after the trip, since all information about that person (right down to the last atom, let's say) would have been transferred in the teleportation beam. But, what if the teleportation signal were duplicated, so that *two* of 'me' find themselves on the alien planet? This seems to me to be the fastest and surest way to make a fresher philosophy student contradict themselves. They end up wanting to say that the two people are both me, because they are both identical to me. But then, since I am only one person, we cannot say that they are two different people. And they clearly *are* two different people!

Thus inconsistency is one of the principal foundations of philosophical pedagogy. The thing about inconsistency is that it is so *annoying*. Students can get really annoyed about examples like the one just given, and will spend time attempting to reconcile their beliefs without thinking of it in any way as 'homework' (perhaps this is the *only* way to ensure that students will do philosophy when they leave the classroom). And not only is inconsistency a key ingredient of philosophical pedagogy, it is also an important part of philosophy itself. Most of the paradoxes in Michael Clark's *Paradoxes from A to Z* (2002), for example, have not been 'solved', and are still discussed in the literature from time to time. It would not be too much to say that if inconsistency were not as important as it is, philosophy itself would not be as important as it is. It has been with us since the very beginning, with relevant passages from both Plato and Aristotle which have been regurgitated and debated ever since.

Of course, most of the inconsistencies that turn up in philosophy generally are not inconsistencies in *science*, but if anything inconsistency is even more important in scientific contexts. For one thing, the likelihood of inconsistency is increased dramatically in science by the presence of mathematics and the very precise statements this gives rise to. Then there is the fact that so much of importance is affected by our scientific picture: technology, policy, religion, etc. With inconsistency comes an immediate, justified loss of confidence, and clearly the consequences of this can be dramatic. Within science itself, when an inconsistency is noticed all experiments designed to test the possible truth of the theory can be stopped: we already know that it is impossible that the theory be true, because it is self-contradictory. At the very least *something* needs to be changed in the theory (at least, if we care about truth). Thus—it would seem—scientists gain a great deal from the discovery of inconsistency: it tells them to put fewer resources into experiments and to spend more time doctoring the theory, or trying to devise a new theory altogether. When their theory is *consistent*, by contrast, they have to go out into the world and test it. If it is true, no amount of testing will ever show this for certain. If it is false it is *sometimes* possible to establish this, but usually it takes a great deal of time and effort to do it convincingly.[1] Thus there is little wonder that inconsistency in science has been an important topic in philosophy for several decades, and one attracting more and more attention in the community.

This gives some flavour of the importance of inconsistency in science, but I should also say something about why I am keen to consider inconsistent scientific *theories*, and the internal inconsistency of theories in particular. The truth is, I'm not keen to consider theories (in a sense to be clarified), but I am keen to consider the literature on inconsistent scientific theories. What is so important about the internal inconsistency of theories? Well, put simply, it might be hard to see how theories can work at all if they are internally inconsistent. In such circumstances we can't believe what the theory says, and it looks like the normal rules of logic can't apply. In addition, the standard 'theories of theories' of philosophers of science seem to break down for inconsistent theories (as we'll see shortly). Little wonder that 'internal consistency' has long been considered an important—even crucial—criterion for any scientific theory to be considered 'good' (cf. Kuhn 1977: 321–2; Newton-Smith 1981: 229).

---

[1] If the theory says that the world is a certain way at *all times*, past present and future, this can (sometimes) be tested, whereas in the case of the criminals' alibi their claim that the world was a certain way *in the past only* cannot be tested absolutely.

The *mutual* inconsistency of theories gets far less attention in the literature, since there is no problem of representation for the individual theories, nor a logical or doxastic problem for these individual theories. Of course, we can't believe *both* theories in such circumstances, but since each theory works perfectly well on its own, this needn't get in the way of the day-to-day practice of scientists.[2] Now, as I've noted, in the end I won't be thinking in terms of 'theories', so the distinction between internal and mutual inconsistency won't arise for me. But one of my main objectives is to provide a critical analysis of the philosophical literature on inconsistency in science, and so I need to examine the same inconsistencies as this literature. And the inconsistencies usually referred to as 'internal' are suitable test cases for me to make the important points I want to make.

But why care about theories at all? Well, there is something of a consensus that scientific theories are of paramount importance as units of analysis: since the very beginning of modern philosophy of science *scientific theory* has been one of the most central concepts. This is both implicit to philosophical practice and something philosophers regularly acknowledge. Consider the following passages from Suppe (1977) and Niiniluoto (1984):

If any problem in the philosophy of science justifiably can be claimed the most central or important, it is that of the nature and structure of scientific theories, including the diverse roles theories play in the scientific enterprise. For theories are the vehicle [sic] of scientific knowledge, and one way or another become involved in most aspects of the scientific enterprise. It is only a slight exaggeration to claim that a philosophy of science is little more than an analysis of theories and their roles in the scientific enterprise. A philosophy of science's analysis of the nature of theories, including their roles in the growth of scientific knowledge, thus is its keystone; and should that analysis prove inadequate, that inadequacy is likely to extend to its account of the remaining aspects of the scientific enterprise and the knowledge it provides. At the very least, it calls for a reassessment of its entire account of scientific knowledge. (Suppe 1977: 3)

The nature and function of scientific theories is perhaps the most central problem within the philosophy of science. (Niiniluoto 1984: 111)

---

[2] Examples of mutual inconsistency include thermodynamics and the theory of evolution (on the prediction of the age of the earth), and unresolved examples such as the conflict between general relativity and quantum theory.

Passages such as these gain support from the community simply by the fact that papers on the nature of scientific theories have been in steady supply for the past several decades (as we will see in detail in Chapter 2). And aside from these papers there are many more where, although the nature of theories is not discussed, *theory* is nevertheless in play as a central concept.

Today it is sometimes commented that in recent decades there has been a move away from theories and towards scientific models and modelling. However, even those at the forefront of analyses in terms of models readily admit that theories are still a fundamental part of philosophy of science, as Morrison's (2007) recent paper 'Where have all the theories gone?' attests. In fact the latter is one of many recent papers to re-acknowledge the importance of 'the theory' to philosophy of science as the pendulum of emphasis swings back from models to theories. And even during the 1980s and 1990s, when there may have been a fall in the number of papers analysing the nature of theories, there were still several discipline-defining debates where *theory* featured as a central concept: think of the under-determination of theories by evidence, the pessimistic induction on past theories, the reduction of one theory to another, inter-theory relationships in general, theory confirmation, the concept of a 'final theory of everything', and so on. Add to this the ongoing use of the concept *theory* in science itself and the ubiquity and apparent importance of theories becomes clear. Little wonder the question 'What is a scientific theory?' is still considered a central question in philosophy of science, as stated explicitly in the recent discipline-defining tome *The Routledge Companion to the Philosophy of Science* (Psillos and Curd 2008: p. xix).

In summary, inconsistency already plays an important role in everyday life, and is particularly important in philosophy and science (or so it would seem). And *scientific theory* is one of our most ubiquitous concepts within science and philosophy of science. Thus there is little wonder that interest in inconsistency in science, and in scientific theories in particular, has enjoyed increasing popularity as a topic for philosophers, especially as more and more examples of inconsistency in science are (apparently) unearthed. However, despite a significant number of papers now devoted to the topic, there are some important underlying misunderstandings which still pervade the literature, as this book sets out to show. Dramatically enough, I will argue that inconsistencies aren't really ubiquitous in science, nor very important when they do turn up. And in addition the concept *theory* is not only not important, and usually not helpful, but often harmful. But I anticipate.

1.2 Inconsistency in Science: Contemporary Debates

Philosophers interested in inconsistency in science draw on a string of examples of apparent inconsistencies in science and the history of science. Many of the examples most commonly referred to will be discussed in this thesis: Bohr's theory of the atom, Newtonian cosmology, the early calculus as introduced by Newton and Leibniz, Aristotle's theory of motion, Olbers' paradox, classical theories of the electron and, most recently, classical electrodynamics. Other famous examples of apparent inconsistency in science are Zeno's paradoxes, Planck's theory of blackbody radiation, and the properties of Dirac's delta function.

In the past sixty years philosophers have made much of this string of apparent inconsistencies in science. At the time of writing, two debates are prominent in the literature: (i) the 'content-driven' versus 'logic-driven' debate, and (ii) the issue of how best to represent/reconstruct inconsistencies in science.

1.2.1 *Logic-driven versus content-driven*

For several decades now two central questions have been,

(1) When faced with an inconsistent theory, what do scientists do?
(2) When faced with an inconsistent theory, what *should* scientists do?

Answering these questions in the philosophical climate of the 1940s and 50s, Karl Popper's answer was to reject such a theory outright. In 1940 he wrote,

> If one were to accept contradictions, then one would have to give up any kind of scientific activity: it would mean a complete breakdown of science. This can be shown by proving that *if two contradictory statements are admitted any statement whatever must be admitted*; for from a couple of contradictory statements any statement whatever can be validly inferred.... A theory which involves a contradiction is therefore entirely useless *as a theory*. (Popper 1940: 408, original emphasis)

In 1959 he reiterated as follows: '[consistency] can be regarded as the first of the requirements to be satisfied by *every* theoretical system, be it empirical or non-empirical' (Popper 1959: §24). Even much later Newton-Smith was to agree with this sentiment (1981: 229). But why exactly should we react so radically when faced with inconsistency? Is Popper right that 'if two contradictory statements are admitted any statement whatever must be admitted'?

The point Popper is making is that, according to classical logic, an inconsistency leads to a contradiction—indeed this is the very definition of inconsistency—and any statement we want follows from a contradiction. Suppose we have a contradiction (with 'A' standing for some proposition):

(i)   A&~A

Then from the conjunction we can infer both conjuncts:

(ii)   A
(iii) ~A

Now from (ii) we may infer,

(iv)   A ∨ B

where B is any arbitrary proposition. This may seem like a suspicious move; why can we do this? Well, obviously if (ii) is true then (iv) *must* be true; in other words the inference is truth-preserving, and this is simply the gold standard for good inferences (see next chapter for more detail here). And we may infer from (iii) and (iv), by disjunctive syllogism,

(v)   B

This demonstrates what is called ECQ, *ex contradictione quodlibet*, which can be roughly translated as 'from a contradiction everything follows'. That is, from a contradiction one can infer, with the primitive inference rules of classical logic, every possible proposition—thus its other name 'logical explosion'. It is important also to note that, starting from (ii) and (iii), logical explosion also follows from explicit contradictories (without first requiring a contradiction). Indeed, in the quotation given above Popper (1940) says exactly this.[3]

---

[3] 'ECQ' is therefore a bit misleading, since it suggests that one *needs* a contradiction for logical explosion. In what follows I use the acronym 'ECQ' to refer to a process of reasoning leading to logical explosion which proceeds via a contradiction *or* explicit contradictories.

8   UNDERSTANDING INCONSISTENT SCIENCE

Since Popper was working within the framework of the syntactic approach of Carnap, Reichenbach, and other logical empiricists who tried to represent scientific theories as deductively closed formal systems, from his point of view any inconsistency led inevitably to such absurdity. But Popper's claims have ever since been used as a foil against which to make quite the opposite claim: that inconsistencies can and do exist in science without destroying the enterprise or even affecting it a great deal. This claim is apparently warranted by looking to the historical record. Lakatos asked in 1966, 'Must all inconsistent theories be ruthlessly eradicated as utterly useless for rational argument?', and answered firmly in the negative, giving 'Frege's logic', 'Dirac's delta function', and 'Leibniz's calculus' as examples (1966: 59). Similarly Feyerabend wrote in 1978,

The objection that a contradiction entails every statement applies to special systems of logic, not to science which handles contradictions in a less simpleminded fashion. (Feyerabend 1978: 154)

Characterizing an inconsistent theory as one 'which, after a few steps gives us p&~p' (1978: 157) his examples are 'Dirac's formulation of elementary quantum theory', 'classical statistical mechanics as developed by Maxwell, Boltzmann and others', 'the older quantum theory', and 'the calculus of fluxions'. Shapere (1984a) makes a similar point:

[W]hatever its advantages, consistency cannot be a requirement which we impose on our ideas and techniques on pain of their rejection if they fail to satisfy it. (Shapere 1984a: 235f.)

This time the examples are the 'Dirac delta function' and the 'infinitesimal calculus'.[4]

If it is agreed that scientists do not, and in fact should not, reject theories when they come across inconsistency, then how should questions (1) and (2) be answered? How do we and how should we handle inconsistent scientific theories? One popular answer has come from developments in logic: one should modify one's logic so that the logical explosion noted above does not materialize. In 1948 Jaśkowski cited the presence of inconsistencies in scientific theories as a motivation for the development of the first paraconsistent logic (Jaśkowski 1948), where Aristotle's treasured law

---

[4] I won't discuss Dirac's delta function in this book. The interested reader is directed to Davey (2003).

of non-contradiction is explicitly rejected along with certain classical rules of inference. The answer to questions (1) and (2) is that scientists either do already (subconsciously perhaps) or *should* adopt one or another paraconsistent logic, so as to continue to reason from the theory without engendering the 'logical explosion' of ECQ (cf. Priest 2002; Bueno 2006). Developing such logics has become a global enterprise. In particular we find a large community of paraconsistent logicians in South America, heavily influenced by Newton da Costa who has developed many such systems since the 1950s (see da Costa and French 1990).

A less developed research programme looks to reject this paraconsistent approach, and to handle inconsistencies in a different way. As a comparison to the logic-driven approach of paraconsistency Norton (2002) has termed the alternative the 'content-driven approach'. The suggestion, roughly, is that inconsistencies in science *are* or *should be* handled by '[reflecting] on the specific content of the physical theory at hand', whilst maintaining classical deductive logic (Norton 2002: 192). Norton's many important papers on inconsistency in science (1987, 1993, 1995, 1999, 2000, 2002) are characteristic of the approach, and Norton cites Smith (1988) as working in the same vein. Frisch (2005a) has recently taken up the fight, criticizing the use of paraconsistent logics and writing,

[E]ven if there were some formal [paraconsistent] framework in which the allowable derivations in a given theory could be reconstructed, any argument for the adequacy of the framework would have to piggyback on an informal 'content-based' assessment of which inferences are licensed by the theory and which are not. (Frisch 2005a: 40)

Some work remains to be done in clarifying just what the content-driven approach consists in, and explaining the differences between Norton, Smith, Frisch, and others. 'Content-driven' sometimes appears to be used to refer to any approach that is not logical. For example, perhaps Harman's approach (1986) can also count as content-driven. He suggests that when faced with an inconsistency one can maintain classical logic and work around it, simply trying to avoid undesirable consequences.

Clearly there is something important about questions (1) and (2): one might imagine that good answers to them could give us clues as to how we might make progress in the face of *current* conflicts, such as between general relativity and quantum theory. But is it right to put such an emphasis on scientific *theories* in these questions? How should we think about theories in the first place? This is a primary concern of the other major debate on the topic of inconsistency in science: the reconstruction/representation of inconsistent theories.

1.2.2 *Reconstructing/representing inconsistencies*

The question of interest here is,

(3) How should we represent/reconstruct inconsistent scientific theories?

The motivation for the question is clear: the two reconstructions of scientific theories most popular in the past sixty years—the syntactic and semantic approaches—both apparently fail as reconstructions of inconsistent scientific theories.

Consider first the syntactic approach: one can reconstruct scientific theories as logically closed axiomatic systems in first order logic, as logical empiricists and positivists suggested in the 1930s, 1940s, and 1950s. But then when faced with an inconsistent theory, if one uses classical logic we get the absurd logical explosion noted above, so that every inconsistent theory just is the set of all possible propositions. Popper was working within this framework so it is little wonder that he was so keen to reject inconsistency. Since the historical record seems to allow for inconsistent scientific theories, one is tempted to reject the syntactic approach on these grounds alone.

However, when the syntactic approach *was* rejected in the 1960s, the main view to replace it similarly didn't seem to be able to accommodate inconsistent theories. The 'semantic' or 'model-theoretic' approach draws on Tarski's conception of a structure which 'satisfies' a set of constraints. As Brown put it in 1992,

Since there are no models of inconsistent sets of sentences, straightforward semantic accounts fail. (Brown 1992: 397)

Two different ways to approach question (3) have been suggested by Nickles (2002) and Frisch (2005a). Nickles' suggestion is to decide how best to reconstruct theories in the first place, and only then consider the inconsistency of theories. The thought is that how we understand theories dramatically affects how we understand their inconsistency, so there is no point in asking questions of the inconsistency of theories before we have an adequate analysis of theories in general. Considering the change of direction from syntactic to semantic characterizations of theories, and what inconsistency looks like on the new picture, he writes,

[O]ur very characterization of science is at stake here, and this in turn largely determines what is at stake when inconsistency arises. (Nickles 2002: 8)

This profound change in the conception of theories and their relation to applications raises several questions about what inconsistency is and how it is to be identified and dealt with in practice. (Nickles 2002: 11)

So it looks like, following the triumph of the semantic approach over the syntactic approach, there are many new questions to ask about inconsistency. Our approach to inconsistency should change because our approach to theories has changed. But then if a change in our concept of a theory makes such a difference it will be very difficult to proceed with confidence. Which version of the semantic view should be preferred? And what if a new analysis of theories comes along? Could it be the case that we ultimately want to reconstruct theories in such a way that they are not even the *kind* of thing which can be inconsistent? Such thoughts suggest that, if we don't want to waste our time, we ought to sort out our disagreements on the nature of scientific theories before we go about discussing their inconsistency.

On the other hand, Frisch (2005a) suggests that we use the fact that many theories *are* inconsistent to decide how best to reconstruct them. He first notes that neither of the two most popular accounts will do:

If theories are identified with deductively closed sets of sentences [syntactic], then consistency is a necessary condition for a set of sentences being an even minimally successful theory (if we assume that inferences from the theory are governed by classical logic). By the same token, if we think of a theory's models as structures in which the theory's laws or axioms are true [semantic], then the laws of the theory need to be consistent. For a theory with inconsistent laws has no models. (Frisch 2005a: 7)

He then goes on to suggest a non-Tarskian semantic approach—along the lines Nancy Cartwright (1983) proposes—and concludes that our conception of theories should be 'model-based' (2005a: 12) such that 'theories do not have a tight deductive structure' (2005a: 11).[5]

---

[5] Cartwright has resisted the label 'semantic' for her approach, but it is usually referred to as 'semantic' because of its focus on scientific models, even if these models are not models in the sense of model-theory.

Frisch's method follows that of da Costa and French (2003), at least to the extent that it starts from the fact that many scientific theories *are* inconsistent. But da Costa and French opt to alter the semantic approach in a different way: for several years they have been developing a modification of the semantic approach—the 'partial structures' approach—which is inconsistency tolerant. They write,

[R]egarding theories in terms of partial structures offers a straightforward and natural way of accommodating inconsistency. (da Costa and French 2003: 85)

Frisch (2005a: 39) criticizes 'partial structures' as essentially just another attempt at a logic-driven approach to inconsistency. If one is going to modify the semantic approach in this way one could equally modify the syntactic approach by swapping the classical closure of the axioms for a nonclassical closure. Of course, the semantic approach is clearly an improvement on the syntactic approach for other reasons, but it would be nice to have an approach which accommodates inconsistency more naturally.[6]

1.3 Going Deeper

Questions (1), (2), and (3) are hardly the only questions about inconsistency in science which might be asked. An obvious question is,

(4) Which scientific theories are inconsistent?

But, even if there were no issues regarding the concept *scientific theory*, this question is hopelessly ambitious. First of all, discovering a new inconsistency is something only those scientists and mathematicians working on the details of a given theory are qualified to do; it is not a task for the philosopher of science. Second, even those most intimate with the details of a given theory cannot hope to set out to prove its consistency or inconsistency. Instead, individuals usually come across inconsistencies quite by accident, whilst in the course of some theoretical investigation (we'll see some examples later on). In fact—given Gödel's second incompleteness theorem—it is impossible to prove the consistency of theories which

---

[6] For my personal take on the 'partial structures' approach to inconsistent science, see Vickers (2009b).

draw on even a modicum of modern mathematics. The inconsistency of a given theory *can* be proved, but even very simple theories can hide inconsistency very well indeed. For example, Smith (2007) considers the theory Q_G, which consists of some extremely basic truisms of mathematics and Goldbach's conjecture, that any even number is the sum of two primes. He writes,

> Q_G is consistent only if the conjecture is true.... But no one knows whether Goldbach's conjecture *is* true; so no one knows whether Q_G is consistent. Which is a powerful reminder that even *very* simple theories may not wear their consistency on their face. (Smith 2007: 67, original emphasis)

Thus the present work takes up the much more modest task of assessing the inconsistency claims which have already been made. This is important, because there is considerable disagreement about whether many of these theories really are inconsistent (as we will see). So the original question I wanted to answer was,

(5) Which of the scientific theories usually labelled 'inconsistent' are genuinely inconsistent?

However, I soon came to regard this as a bad question, or at least an unhelpful question. The fact is that the two main concepts at issue, *inconsistency* and *scientific theory*, are often handled in a very intuitive way by philosophers of science. That is, philosophers of science only very rarely identify an inconsistency in a theory in a rigorous or even semi-rigorous way, and they often speak of theories in general or specific individual theories without giving any details about how exactly they would identify the content of a theory. This makes claims vague, and leaves lots of room for miscommunication as philosophers use the same terms to refer in slightly different ways. Thus the first task of the current thesis is to get absolutely clear on how we should use the terms 'inconsistent' and (especially) 'scientific theory'. Getting clear on these two terms will be the task of Chapter 2.

Why does getting clear on the key concepts make question (5) a bad question? Well, (5) implicitly makes certain claims, about scientific theories in particular, which might not be true and which anyway don't need to be made. In fact, all five questions (1)–(5) assume that scientific theories both (i) are the *kind* of thing which can be inconsistent, and (ii) have definitive, canonical content, such that there is a fact of the matter as to whether they are inconsistent or not. But the former assumption is false if (say) theories are families of models, as many philosophers have claimed in recent decades. And the latter assumption is false if (say) we should be pluralists about theories, or about theory content, or if theory membership is in a sense vague or fuzzy.

As a matter of fact philosophers just do disagree about theories, on just about every point one can imagine (as we'll see in Chapter 2). So we might think that the people who are right about which theories are inconsistent are those that have the 'right' conception of theories. I have come to doubt that there is such a thing as the 'right' conception. Originally I set out to argue precisely this. But then I came to realize that I could properly assess the claims about inconsistent theories which have been made without taking any stance on the nature of theories (including whether there is a right conception or not). How is this possible? Well, one simply looks at the propositions which have been (or need to be) put together to reach inconsistency, and considers why (if) that particular inconsistency is an interesting or important one. For example, does it tell us something about how science works, or could work? This can all be done without even addressing the question of whether or not the set of propositions in question deserves the label 'theory'.

With this new approach in hand a fresh look at the classic cases of 'inconsistent scientific theories' is warranted. What is required is an appreciation of the character of the propositions which must be put together in a given context to reach inconsistency, and what the inconsistency teaches us given that character. Thus, in Chapters 3 through 7, I turn to several classic examples of 'inconsistent scientific theories' and how each fares on the new analysis. If one insists on using the word 'theory', the closest one gets to my question is,

> (6) How *must* the content of these theories be identified if they are to come out inconsistent, and what do we learn given this identification?

But this still isn't ideal, since many will no doubt object that the sets of propositions in question aren't theories at all, and may even reject the analysis on that basis. This response would be a mistake, since it really doesn't matter whether we do or don't have 'a theory': what matters is whether there is something important to be learnt from the fact that the particular set of propositions in question is inconsistent. Thus the real question of interest here will be,

(7) In each case, which propositions have to be put together to reach inconsistency, and what do we learn given the character of those propositions, their relationships with each other and other propositions, and their relationship with the real history of science?

Although many of the cases in question might be referred to as 'theories', there is no need to use this word and perpetuate the controversies and miscommunications to which it so often gives rise (as we will see vividly in later chapters).

How then does this thesis relate to the debates noted above? It is certainly not independent of them. In many respects it is more fundamental: it gets right to the heart of what matters about inconsistency in science whilst avoiding unnecessary side-issues concerning the nature of theories. Take, for example, the two most debated questions, (1) and (2). These both assume (or at least suggest) that theories will all react to inconsistency in the same way, so that scientists will/should all react to inconsistent theories in the same way. But if instead we leave theory-talk out, and ask question (7), then it leaves open the possibility that some of the things we call 'theories' will have very different characteristics, such that inconsistency means something different in each case. (This, in fact, is what we will find in the forthcoming chapters.)

Question (3) is also misguided. If we really can ask and answer all of the truly important questions about inconsistency in science without referring to theories, then it is no longer relevant how theories (consistent or not) are represented/reconstructed. Whether theories should be represented as families of models or not (or whether this is a bad question) makes no difference to whether or not there are sets of propositions, relevant to science and/or the history of science, the inconsistency of which teaches us valuable lessons about how science works.

Question (4), like question (3), makes an assumption that there is a way theories are, or should be represented. Then, like questions (1) and (2) it assumes (or at least suggests) that theories are all basically the same kind of thing, such that inconsistency means the same thing for each one. Otherwise finding out that a theory is inconsistent would be pointless, because we wouldn't know what this meant without also knowing the sense in which it is 'a theory'.

So in the end questions (1)–(6) either (i) are based on premises which turn out to be false, or (ii) at least make assumptions about theories which are controversial, which may indeed be false, and which we don't need to make to learn the important lessons about inconsistency in science. Better to go straight to the heart of the matter, simply identify the propositions which *are* inconsistent in a given case, and investigate whether there is something interesting or important to be learnt from that. This, roughly speaking, will be the approach. I will argue for it and explicate it in detail in Chapter 2. This will be combined with some considerations regarding the concept *inconsistent*. Under precisely which circumstances is a set of propositions inconsistent? It turns out there is more to this than the definitions in the logic textbooks.

2

Concepts and Method

2.1 Introduction

It is prudent to clarify the key concepts of one's subject matter before one launches into the analysis. In many cases this ends up simply being a matter of a recap on current thinking about a given concept, or perhaps a brief argument for why one chooses to use one conceptual framework as opposed to another, perhaps hinting at the problems that will be solved using that framework. For the present project, an investigation of the state of play vis-à-vis the concept *scientific theory* will lead to a radically new method for understanding inconsistency in science. And this is not just a new method to be compared with all the others available, but rather something altogether more fundamental. But first, I turn to the concept *inconsistency*, which is a simpler affair. Although not as simple as most logic textbooks would have us believe.

2.2 What is Inconsistency?

If we're in the business of assessing claims that certain scientific theories are inconsistent, then we ought to be very clear on precisely when it is appropriate to use the word 'inconsistent'. Here the obvious move is to draw on standard definitions provided by the logician. One typically finds that a set of statements are inconsistent iff: (1) a contradiction can be derived, or (2) there is no interpretation of the statements such that they all come out true. For example, consider the following three statements:

(a) Paul is married.
(b) Paul does *not* have a spouse.
(c) If Paul is married, Paul has a spouse.

A contradiction 'A&~A' can be derived using standard rules of inference: using *modus ponens* we have it from (a) and (c) that ~(b). So from (a), (b), and (c), using the inference rules *modus ponens* and *&-introduction*, we reach 'A&~A' where 'A' stands for 'Paul has a spouse'. In addition there is no interpretation of the statements such that they all come out true. By this we mean, *however* one interprets the non-logical terms 'Paul', 'married', and 'spouse', one will never be able to make (a), (b), and (c) all come out true. We still have inconsistency if we strip these three non-logical terms of meaning: the statements are inconsistent by virtue of their logical form alone.

A complication that is sometimes noted is that the inconsistency of a set of statements is relative to a set of 'allowed' inference rules. But there is a standard set of rules judged to be truth-preserving, including *modus ponens*, *modus tollens, disjunctive syllogism*, and a few others.[1] Or, to put this another way, if one starts with a set of propositions Δ, and if these statements are all true, then any use made of the standard rules (syntactic entailment '$\vdash$') will lead us to other true statements (semantic entailment '$\vDash$'). When Δ is inconsistent one might then reason as follows:

(i) $\Delta \vdash$ A&~A (for some A)
(ii) So $\Delta \vDash$ A&~A (the logic is sound)
(iii) So if the statements in Δ are all true, then 'A&~A' is true
(iv) But 'A&~A' cannot be true (it's a contradiction!)
(v) So at least one statement in Δ is false

Some philosophers see this as the end of the matter. Others wish to make something of the fact that this analysis completely ignores any meaning of non-logical terms. Take the first two propositions used above:

(a) Paul is married.
(b) Paul does *not* have a spouse.

According to the above definitions, these two propositions are consistent. There is certainly an interpretation of the non-logical terms such that the

---

[1] In fact there has been some debate about the truth-preservation of these standard rules. It has been shown that we can take *modus ponens* to be the sole inferential rule of classical logic (see Hunter 1971), so the question reduces to whether *modus ponens* is truth preserving. A few have argued that it is not, but I will side with Salmon (1965) who mounts a typical defence.

propositions come out true (for a particular Paul). For example, if we interpret the word 'married' to mean *single*, then we have consistency. And one can't derive a contradiction from these statements, at least using any set of inference rules which might be called 'logical'. There is no rule which allows us to move from 'married' to 'has a spouse'.

One cannot put too much weight on the word 'logical' here: it is well documented that there is no precise cut-off from the 'logical' to the 'non-logical'.[2] Some have argued that meaning is prior to rules, that the logical rules are decided by the meaning of terms such as 'and' and 'or', and that there is no reason to hold back from adding new rules to *modus ponens*, *disjunctive syllogism*, and the rest if a case can be made that an inference is truth-preserving (cf. Brandom 1994). So if the inference from 'married' to 'has a spouse' is truth-preserving, then one might claim that (a) and (b), above, are inconsistent, and that we don't need (c) to reach inconsistency. The definition of inconsistency might then go as follows:

> A set of propositions Δ is inconsistent *iff* a contradiction follows by truth-preserving inferences of *any* kind.

In this way, no restriction to logical inferences is made, and so-called *material* inferences might be introduced (Kapitan 1982; Read 1994; Brigandt 2010).

These sorts of considerations can lead to some very thorny issues concerning meaning, conceptual/semantic content, the nature of propositions, and so on. I am pleased to report that most of these issues will not matter here. The reason why will become clearer once I have introduced other aspects of my method. Accordingly I now turn to the concept *scientific theory*; I will return briefly to inconsistency in §2.7.

2.3 Theories of Theories: Syntactic and Semantic

I have already mentioned that I am ultimately going to do without the concept *theory*. I could just outline and adopt the method, and then go on to apply it in the case studies, justifying its use by the results achieved. However, for many this will look like a radical method without sufficient

---

[2] See e.g. Brandom (1994: 94ff.), Lycan (1994: 243), Warmbrod (1999), MacFarlane (2000).

motivation. Accordingly, in the next three sections I will outline some of the problems we would face if we made use of one or another 'theory of theories', or one or another conceptual analysis of *theory*. (If the reader is happy to take my word for it on this issue, he/she may jump directly to §2.6.)

Analyses of *scientific theory* are ubiquitous in the history of philosophy of science. The first systematic, detailed analyses were the syntactic analyses of Carnap (1939, 1950, 1967), Reichenbach (1951), Braithwaite (1953), Nagel (1961), and others. These were followed up by the semantic approaches of Suppes (1967), Suppe (1977, 1989), van Fraassen (1980), Giere (1988), da Costa and French (2003), and others. But beside these two major research programmes many other suggestions have been made (although it is often left ambiguous as to whether an analysis of *theory* is intended, or whether a replacement for 'theories' is intended). Two of the more famous suggestions are Kuhn's 'paradigms' (1962) and Lakatos's 'research programmes' (1970a). Alongside these well-known analyses one can point to Heisenberg's 'closed theories' (see, e.g., Bokulich 2006), Neurath's 'theory balloons' (see, e.g., Cartwright et al. 1996), Finkelstein's 'three levels of theory' (1966), Toulmin's 'conceptual systems' (1972), Shapere's analysis of theories in terms of 'domains' (1977), Laudan's 'research traditions' (1977), Gould's 'essence' approach to evolutionary theory, and his 'Goldilocks' approach to the concept of a theory (2002), Wilson's 'theory façades' (2006), Morrison's 'theoretical core' (2007), Hendry and Psillos's 'consortia of different representational media' (2007), Darrigol's 'modular' conception of scientific theories (2008), French's 'quietist' approach (2008), and the recent 'Hierarchical Bayesian Perspective' (Henderson et al. 2010). In addition there are many papers which call for unappreciated 'types' of theory, such as Finkelstein's 'categorical' and 'flexible' theories (1966), Darden and Maull's 'interfield theories' (1977), Rohrlich and Hardin's 'established theories' (1983), Batterman's 'theories between theories' (1995), and Craver's 'mechanistic theories' (2002).

Many of the accounts of theories and theory-types in the books and papers just noted are complex, and there are complicated and unclear overlaps and interrelations between the accounts. What is one to do when faced with this wall of opinion? Can we ever get to the 'true nature' of the scientific theory? What can be done, aside from choosing from the start one conception of theory over all the others, trying to justify that choice as best one can?

Today any discussion of the question 'What is a scientific theory?' invariably starts by comparing the syntactic 'received view' approach and the semantic ('model-theoretic') approach. These are two very different approaches, but as Morrison has recently put it, 'What these views do have in common, however, is the goal of defining what a theory is' (2007: 198). Chakravartty (2001) starts with the question 'What, precisely, is a scientific theory?' and then goes on to discuss how on the syntactic view 'a theory is an axiomatic system' whereas on the semantic view 'a theory is a family of models'. Da Costa and French (2003) also follow this trend, introducing the syntactic view as follows,

> Let us return to the issue—which we regard as absolutely fundamental—expressed by the question 'what is a scientific theory?' According to the so-called Received View ... the answer is relatively straightforward: A theory is an axiomatic calculus given a partial observational interpretation via a set of correspondence rules; that is, a theory on this view is a logico-linguistic entity. (Da Costa and French 2003: 23)

In other words a theory is meant to 'be' a set of axioms written in first-order logic. Primitive, non-logical terms are interpreted by their logical relationships with each other and rules for how they 'correspond' to certain observation terms, that is, terms which in turn correspond to observed phenomena.[3]

The downfalls of such an attitude are now well documented, and the approach has not been taken seriously by those investigating the nature of theories since the 1970s. Since that time the most popular conception of theories has been 'semantic'. Writing as the tide was turning, Suppes' paper 'What is a scientific theory?' (1967) essentially says that theories are semantic, not syntactic; in other words that one should focus on models—in the sense of Tarski's model-theory—rather than the 'logico-linguistic' axioms of the syntactic approach.[4]

There are complicated stories to tell about both of these approaches, but this is not the place to do so. The crucial point here is that, on a perfectly

---

[3] For more detail, see for example, Craver (2002) and French (2008). For a definitive discussion, see Suppe (1977).

[4] A more detailed analysis of Suppes' (1967) paper sees it as discussing the *representation* of theories, rather than the question of what they *are* (see, for example, French and Saatsi 2006: 552–3). If this is right then it is unfortunate that Suppes did not entitle the paper 'How Should Scientific Theories be Represented?' or something similar.

reasonable reading of the literature, 'what a scientific theory is' is something very different on the two accounts. Indeed 'what a scientific theory is' can vary dramatically across the variants of just one of these accounts. Now recall what was noted in §1.1, that there are several discipline-defining debates in philosophy of science where the 'scientific theory' plays a central role. Little wonder, then, that Suppe referred to the scientific theory as the 'keystone' of philosophy of science. Because it looks like how we approach any one of these debates is going to be very different depending on how we choose to characterize our theories, and the conclusions of these debates will likewise be very different. Thus, as Suppe says, when a philosophy of science gets theories wrong, 'it calls for a reassessment of its entire account of scientific knowledge'.

This is where the representing/reconstructing debate noted in Chapter 1 comes in. Nickles (2002) says that we should decide what theories are first and only *then* tackle inconsistency, and Frisch says that deciding what theories are should be informed *by* inconsistency. However, what isn't made clear enough is the difference between asking *what theories are* and asking *how theories should be represented/reconstructed*. The fact that the question 'What is a scientific theory?' has been associated with the syntactic and semantic approaches suggests that they belong to the former question, but their character suggests otherwise. Their character suggests that they are just two different ways to represent/reconstruct science, which in turn suggests that they do not have to be competitors.

Following this line of thought, one might think that instead of picking the 'right' view of theories, we need to think about what the most appropriate view of theories is *for the investigation at hand*. But on the other hand, if we are persuaded that both syntactic and semantic approaches are reconstructions *of* theories, we might wonder whether it is possible to conduct an investigation in terms of the 'theories themselves', the things that the representations are representations *of*.

2.4 The Substrate of Science

If we want to investigate whether scientific theories are *really* inconsistent or not then we might be interested in an account of 'theories themselves', which are more fundamental than the syntactic and semantic reconstructions *of* theories. Such a gap between actual science and the syntactic and semantic views is described in the following way by Hendry and Psillos (2007):

> Both standard views [syntactic and semantic] have been comrades in their attempts to rationally reconstruct scientific theories. Where they differ is in the tools they use. . . . [T]he standard views have alike aimed at *rational reconstruction*.
>
> We do not want to doubt the usefulness of (moderate) formalization and reconstruction. But we should not lose sight of the fact that they *are* reconstructions, or mistake their products for the theories themselves. (Hendry and Psillos 2007: 159, original emphasis)

Does this notion of a 'theory itself' really make sense? Can it be a useful notion? It certainly does not originate with Hendry and Psillos: it can be found in much of the relevant literature, including Giere (1988), Chakravartty (2001), French and Saatsi (2006), and more recently Votsis (2011). But Hendry and Psillos do give the most illuminating account to date as to what exactly 'theories themselves' could be. They go on:

> In pursuing their claims about particular scientific theories, philosophers and historians of science take as substrate the written, drawn and spoken products of scientific theorizing and distil the joint 'content' of these products, using whatever formal tools are available to them. Of course the relationship between substrate and product in this process is not a simple one: the scientists' claims may have to be rounded out to capture a theory's commonly agreed implications, 'filled in' with extra structure for completeness, or cleaned up in the interest of consistency.[5] The nature and extent of this rational reconstruction will reflect the (philosophical or historical) purposes of the analysis. (Hendry and Psillos 2007: 159–60)

The basic idea is that the 'theories themselves' exist within this 'substrate', which is really just what one finds recorded in the annals of science. Philosophers then may have various reasons for reconstructing certain elements found in the annals of science. These reconstructions are sometimes called 'theories', but strictly speaking they are reconstructions *of* theories, and the 'theories themselves' are in the substrate.

---

[5] 'Consistent' here is meant in the sense of 'stable, constant over time and place', rather than 'not self-contradictory'.

Now, in the annals of science what we really find are sentences, diagrams, etc. written down. Ultimately these are just lines on pages. We need to interpret them to reach anything that could be called a theory: on any account a theory is much more than just lines and symbols. Of course, we can look to the annals of science and interpret things naturally, as seems fit to us. But we can get things wrong: we can end up thinking a past scientist thought something that he/she actually did not think. In such a case what we may think is a theory held in the history of science just never existed, and thus it would almost certainly be of no interest to find it to be inconsistent. Thus one might add to Hendry and Psillos's account a hidden 'propositional realm' that we don't find in the substrate, but have to infer *from* the substrate using whatever historical tools we may have at our disposal. Ultimately, it is these propositions, believed, accepted, or entertained by genuine historical actors, that we care about.

With this extra component we might represent things diagrammatically. Figure 2.1 shows just two different types of thing which can be taken from

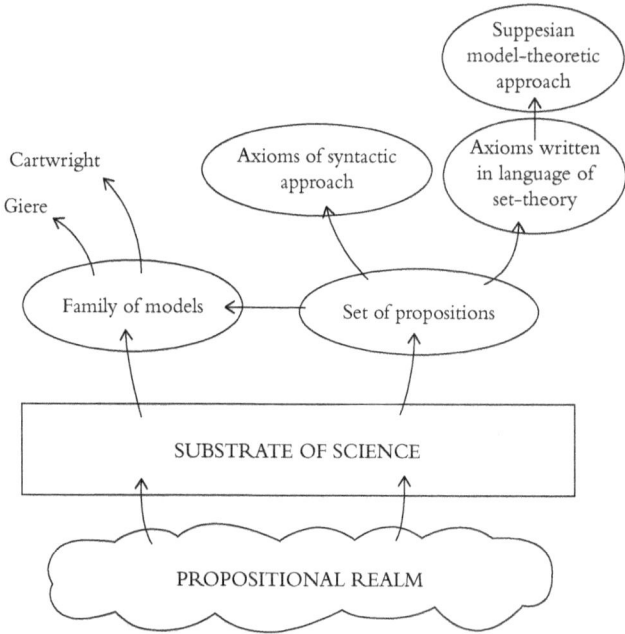

Figure 2.1 The relationship between science and some popular philosophical reconstructions of science

the substrate of science in the context of an analysis of the 'scientific theory': scientific models and propositions (inferred from sentences). The models may come straight from the substrate (as Giere and Cartwright seem to prefer), or they may come via a set of propositions (constraints) which they satisfy. On the other hand one can go elsewhere with a set of propositions, formalizing them syntactically or semantically. Thus we find several different possibilities for what is sometimes called 'reconstructing scientific theories'. One can think of different 'strengths' of reconstruction, in that some reconstructions look like nothing you would expect to find in the substrate (strong reconstructions, e.g. the syntactic approach of the positivists) and others can be almost identical with what is found in the substrate (weak reconstructions). And one can think of the stronger reconstructions as utilizing weaker reconstructions as stepping stones, for example when Suppes starts with a set of propositions and goes on to formulate a set-theoretical predicate (Suppes 1957: ch. 12). Thus in the figure it becomes clear what Hendry and Psillos (2007: 159) mean by both the 'nature' of reconstruction (in which *direction* one travels from the substrate) and the 'extent' of reconstruction (how *far* one travels from the substrate).

This may be all well and good, but what of 'theories themselves'? We can't access the 'propositional realm', and the substrate is ultimately just lines (mostly making up sentences) on paper, so it does seem that we must do at least *some* reconstructive work in our analysis. If we want to do justice to the real history of science, then one lesson might be that we should stay as close to the substrate as possible, and do our best to reconstruct the relevant propositions behind the sentences. But the main issue for the present investigation is how we pick out the *content* of a given theory, because this is what will determine whether a theory is consistent or inconsistent. We are no closer to identifying this content.

2.5 Concepts and Contents of Theories

One option for picking out the content of 'theories themselves' might be to focus on what scientists themselves referred to as 'the theory' in relevant literature. On the face of it there are cases where this seems to work out OK. For example, with Bohr's theory of the atom we find that a collection of propositions are grouped together and labelled 'a theory' within the science, as in the final pages of Bohr's famous trilogy of papers of 1913 (1913a). What could be more obvious than identifying the theory with these propositions? But as we will see in the next chapter, things are not so simple. What people usually refer to as 'Bohr's theory' today is *not* the same thing that Bohr presented as 'the theory' in his 1913 articles. And this isn't just a failure to know one's history: even Bohr's contemporaries didn't think of the theory as constituted by Bohr's 1913 postulates (see Chapter 3).

So the most obvious cases can be difficult, but in many other cases it seems clear that there is no 'theory itself' in the substrate of science and that philosophers later come along and reconstruct a 'theory' from what they find. A good example here is John Norton's (1993) analysis of what he calls 'Newtonian cosmology', as we will see in detail in Chapter 5. Quite obviously in this case there is no textbook or article in the history of science which sets out 'the theory'.

Thus we need something more to pick out the content of a theory. Conceptual analysis is an obvious tool here. If we could agree, for example, that 'T is a theory if and only if T is a proposed explanation of a certain specified phenomenon or domain of phenomena', then we would have something to go on vis-à-vis theory-content. But there are two major problems here: (i) the so-called 'classical' theory of concepts whereby one identifies necessary and sufficient conditions is today buried beneath a plethora of serious objections, and (ii) even if we could hit upon a correct definition of 'theory', disagreements about the *content* that should be ascribed to a given theory would remain.

Drawing on contemporary theories of concepts is little help here. For one thing, there is no easy way to justify picking one such theory over the others. But even if we did pick one such theory (e.g. prototype theory), we would still have the problems of (a) actually providing the conceptual analysis of *theory*, and (b) using that conceptual analysis to identify the content of a given theory in order to determine whether it is inconsistent or not.

Of course, one might claim that we *just do* agree on at least some of the content of any given theory. Newtonian mechanics includes Newton's laws, and classical electromagnetism includes Maxwell's equations. Gould (2002) and Morrison (2007) write of an 'essence' or 'core' of a theory: this is the 'absolutely minimal set of truly necessary propositions' (Gould 2002: 15) or 'a set of fundamental assumptions that constitute the basic content of the theory' (Morrison 2007: 197).[6] Gould's example is 'evolutionary theory', the essence of which is constituted by 'overproduction of offspring, variation, and heritability' (Gould 2002: 13). However, suppose we asked Gould whether evolutionary theory is inconsistent. The fact that there is no inconsistency in the 'essence' of the theory would be little comfort, since this does not constitute the whole theory. In addition, disagreements will remain about just which parts of a theory are 'truly necessary propositions' or 'fundamental assumptions'. It might be easy enough to say that 'classical electrodynamics' definitely does contain Maxwell's equations, but in precisely what form and with what interpretation beyond the bare mathematics? This is an issue that will be fundamentally important in Chapter 4.[7]

I think Gould and Morrison are heading in the right direction. The advantage of their approach is that when the word 'theory' is used just to put us in mind of the 'absolutely necessary' parts, there will be considerably less disagreement than if we are trying to *identify* theories. How much we can do with these necessary parts, and how we introduce other theoretical constituents, is another matter. But in the meantime the concept *theory* is being asked to do far less than usual. As Mark Wilson has put it, '[W]e commonly expect "concepts" to carry great evaluative burdens, yet not buckle under the freight' (2006: 87). *Scientific theory* is just such a concept, a concept that has been asked to do much more than it can manage. The result is rampant cross-talk, miscommunication, and disagreement in the literature, as indicated above and as we will see in the case studies.

What good is this 'minimal' approach for the purpose of deciding whether given scientific theories are really inconsistent? Well, not much, but the point is that the question whether a theory is inconsistent might not be a good question. The disagreement about theories might be symptomatic of there being no fact of the matter about how exactly a given theory 'should' be identified.[8] Instead we are encouraged to think of the 'core' or

---

[6] This is analogous to the neoclassical theory of concepts, where one identifies only necessary conditions, and not sufficient conditions. Indeed, there is an interesting, undeveloped parallel between theories of concepts and stances which might be taken on theory-content.

[7] Another example is the apparent difference between Gould (2002) and others who assume that 'selective retention' is a fundamental part of evolutionary theory. Cf. Burian (1977: 30).

[8] Burian (1977) was ahead of the game here. For example he writes, 'Newtonian mechanics cannot properly be identified with any of its formulations, not even Newton's' (p.37).

'essence' of a theory, plus other theory-elements which might be added to that core for particular purposes.

As I said, this is going in the right direction, but it doesn't go far enough. There will still be disagreement about which are the 'truly necessary' or 'fundamental' assumptions. And if we're in the business of discussing inconsistency in science, we'll still need to add other propositions to this 'core' to reach inconsistency in most (perhaps all) cases. But if we're taking this approach, why burden ourselves with this distinction between core propositions and other propositions? Why not simply talk about sets of propositions?

2.6 A New Method: Theory Eliminativism

I've said that 'theories' play a major role in philosophy of science, and discussions of inconsistency in science are no exception. In this book we will be looking at a number of 'scientific theories' judged by at least some philosophers to be internally inconsistent theories, and we will be assessing those claims. So it may look like we will have to have *some* take on what theories are, and in particular how we should decide upon the content of a given theory.

In fact we can move forward without any such take on theories, and without sacrificing anything of real importance. As already noted in §1.3, my question of interest was never 'Is this theory inconsistent?', but instead 'Does the claim that this theory is inconsistent *as given in the literature* hold up?' But any inconsistency claim made in the literature refers, explicitly or implicitly, to a set of propositions (or 'statements' if you prefer).[9] This is so because by far the most natural referent of an inconsistency claim just is a set of propositions (broadly construed). Now, once we have identified the relevant set of propositions we can ask and answer all sorts of questions without making any use of the concept *theory*:

Q1. Is the set of propositions really inconsistent?

---

[9] For example, Frisch (2005a) prefers a models-based, Cartwrightian theory of theories, and not an account in terms of propositions. But he nevertheless presents his inconsistency in terms of a set of propositions. More on this in Chapter 4.

Q2. In what sense are the propositions historically relevant? What evidence is there that these propositions played a genuine role in the relevant history of science?

Q3. What propositional attitude(s) did the relevant scientific community have to the propositions in question?

Q4. How did/do scientists reason with and draw inferences from the propositions? How was/is logical explosion avoided?

Q5. How do the propositions tie together? What scientific work do they do *as a group*?

Q6. What is the relationship between the logical, mathematical, and physical content of the propositions?

But once one has asked and answered all of these questions, what point remains in asking the question 'Do these propositions really constitute a *theory*?' or 'Do these propositions really constitute [insert theory-name here]?'?

If we're no longer using the concept *theory* it's clear that much is gained: we no longer have to worry about the multitude of issues concerning theories of theories and the concept *theory* discussed above. And disagreements in the community about the concept *theory* can no longer possibly compromise these other debates about inconsistency in science. Further, in eliminating *theory* one is forced to state more explicitly why one's point is an interesting or important point: one can no longer hide behind the word 'theory', suggesting that it is obvious that the point is an important one because it is *about the theory*. This in turn brings one to engage with the history of science and science itself in a more serious way than is usual. This because it is the scientific characteristics and historical relevance of the propositions in question which will make the inconsistency interesting or important. (These benefits are largely promissory notes at the moment: it is the case studies which will bring them to life.)

If we eliminate theory-talk, has anything at all been lost? We can no longer ask whether the set of propositions in question is *really the theory*, but why would we care about this question? Suppose we would care because we want to know whether the set of propositions was really believed, or used, or 'entertained' by the relevant historical actors. Well, one can directly ask whether the propositions were really believed, or used, or entertained, and just leave theory-talk out. Indeed, *whatever* answer is given to the question 'Why would we care about whether the set of propositions in question is *really the theory*?', so long as the answer doesn't refer to 'the theory', then we can use that answer to construct questions which we can ask without making use of the concept *theory*.

One possible exception is 'We want to know whether what the historical actors themselves referred to as "the theory" was inconsistent'. But again, why is this an important question to ask? If it's because that tells us which propositions (or other theoretical constituents) they believed, accepted, entertained, used, etc., then once again we can focus on the latter concepts in place of *theory*. On the other hand one might simply insist that it's intrinsically interesting if the thing historical actors referred to as 'the theory' was inconsistent. I don't share this intuition, and I don't see how it could be justified: it looks to me like a way to simply avoid answering the question. Of course, if a community used some term—for example 'electrodynamics'—to refer to a set of propositions, then that shows that those propositions were in some sense 'in play' scientifically speaking, and were (perhaps) epistemically committed to in some way, or used together in some way. But again, if one considers which questions can be asked without making use of the concept *theory*, one sees that everything is covered. Just how exactly were the propositions in question 'in play' scientifically? Just how were they committed to by the relevant community? Exactly how were they used? Once these questions are answered, what does it matter whether the propositions were or were not referred to as 'the theory' or given some name such as 'electrodynamics'? In addition—as we will see in some of the case studies—individual historical actors and (*a fortiori*) scientific communities often don't have a clear sense of how exactly they intend the term 'theory' or theory-names to refer. Scientists don't have a clearer sense than philosophers of science here; quite the opposite.

This doesn't completely rule out use of the word 'theory'. For example, it may get tedious to keep referring to 'the set of propositions in question', and we may instead wish to call them 'a theory' or give them a theory-name as shorthand. In such circumstances we would be using the word 'theory', but not putting any conceptual weight on it. In fact this approach has been used before: from time to time philosophers decide not to put much weight on the concept *theory* and refer instead to a set of constituents. Mathias Frisch has recently provided an example of such practice in the context of his claim that 'the theory of classical electrodynamics' is inconsistent (as we will see in detail in Chapter 4). What is crucial is to recognize what Frisch means by 'theory'. He writes,

> Throughout my discussion I will refer to the scheme used to model classical particle-field phenomena as a 'theory'. (Frisch 2005a: 26)

He is then relatively clear as to what belongs to this 'scheme'. His claim that 'the theory of classical electrodynamics' is inconsistent should be understood simply as the claim that the constituents of the 'scheme' in question are inconsistent. Whether or not this is an important result can then be decided not by asking whether the scheme is 'really the theory' (Muller 2007; Belot 2007), but simply by looking to the properties of the set of constituents in question. Indeed, asking whether it is 'really the theory' causes crucial miscommunications as each protagonist proceeds with a different conception of both (a) what *the* theory is, and (b) what *a* theory is.

Unfortunately, as the Frisch case shows, using the word 'theory' and theory-names merely as labels doesn't work very well. Readers assume that the word has been used not merely as a label, but to make certain implicit claims about the character of the set of theoretical constituents in question. Thus Belot and Muller object that Frisch does not have the right conception of 'what the theory is'. If we are going to use a word as a label for a set of propositions, it will be better to use a different word. Occasionally I will use the locution 'pointedly grouped propositions' (for those who like acronyms: 'PGPs'). The reason is simple: the inconsistency of a set of propositions will only be interesting and/or important if there is some point to grouping those propositions together (that is, some point beyond the fact that they are inconsistent). It may be that the propositions were used together, or believed to be candidates for the explanatory truth of a given domain of phenomena by a relevant community, or come together as (part of) an answer to a specific scientific question, etc. As we will see in the case studies to follow, the 'point of grouping' of a given set of propositions will be crucial in our assessment of the significance of the inconsistency from the point of view of philosophy of science.

Let me tie up one loose end here. I will inevitably make *some* use of the word 'theory' and theory-names such as 'classical electrodynamics' in what follows. The reason is that I am going to be engaging with a large body of literature that uses such terms quite regularly. In addition it is usually harmless—provided one is careful—to use a locution such as 'Bohr's theory of the atom' just to put the reader in mind of a certain episode in the history of science. This will be the sole purpose of my chapter titles. Absolutely no claim will be made about what the nature or structure of the relevant 'theories' is. But most crucially, any assumptions made by the reader upon reading the words 'Bohr's theory of the atom' (say) will be irrelevant to the analysis. No conclusions pertaining to inconsistency in science will make reference to 'Bohr's theory' or 'the theory' or anything of the kind. Indeed, aside from the title of each chapter, barely any use will be made of such terms. My view is that the concept *theory* and theory-names can be helpful tools of communication in certain circumstances, but (usually) not as elements of a serious philosophical analysis.[10]

2.7 Back to Inconsistency

Let me now return to the issues raised in §2.2 regarding a possible generalization of inconsistency beyond the purely logical, to embrace meaning, or conceptual/semantic content of propositions. I said that we would be able to avoid issues concerning meaning. But, as we will see in some of the forthcoming case studies, the presence of an inconsistency in science can hang on the subtlest nuance of meaning. Or at least it is natural to speak in that way. So won't we have to have an account of how a set of propositions can be inconsistent in virtue of semantic content, and not merely in virtue of logical form?

Lakatos gave this issue some thought in 1970. He writes,

Two propositions are inconsistent if their conjunction has no model, that is, there is no interpretation of their descriptive [non-logical] terms in which the conjunction is true. But in informal discourse we use more formative terms than in formal discourse: some descriptive terms are given a fixed interpretation. In this informal sense two propositions may be (weakly) inconsistent given the standard interpretations of some characteristic terms even if formally, in some unintended interpretation, they may be consistent. For instance, the first theories of electron

---

[10] In Vickers (forthcoming), the foregoing discussion is considerably extended to argue that we should expect the benefits of a theory-eliminating debate-reformulation to heavily outweigh the costs for a highly significant number of debates. Inconsistency in science is just one possible application. See also §8.3, this volume.

spin were inconsistent with the special theory of relativity if 'spin' was given its ('strong') standard interpretation and thereby treated as a formative term; but the inconsistency disappears if 'spin' is treated as an uninterpreted descriptive term. (Lakatos 1970a: 143, fn.3)

Now, in light of this example allow me to sketch two opposing views. The advocate of 'material inference' (Kapitan 1982; Read 1994; Brigandt 2010) will say that there isn't really 'weak' and 'strong' inconsistency. Instead, inconsistency just means that what is being stated is impossible, and this does not come in 'weak' and 'strong' forms. On this view, when Ralph Kronig said 'electrons spin' in 1925 this was indeed inconsistent with relativity. The reason is that Kronig's statement has unambiguous semantic content, and certain conflicts with relativity follow from this content. The conflict follows from the material content of Kronig's statement, and does not depend on any logical form.

But many have opposed this notion of 'material content'. On this opposing view the proposition 'electrons spin' does not entail that parts of the electron travel at certain speeds just by itself, strictly speaking. To really make this inference—and thus to reach inconsistency—one needs to add an implicit premise linking the concepts *spin* and *speed* (just as in the marriage case of §2.2 an implicit premise is needed linking *married* and *spouse*). The reason is that the proposition 'electrons spin' does not somehow contain conceptual content linking *spin* with *speed* all by itself: it only ends up with the meaning it does because of extra implicit premises, so widely and deeply held by the members of the relevant linguistic community that we don't even notice them. In the end, the only inconsistency is logical inconsistency.

Now, my claim is going to be that it won't matter for my purposes which attitude I take to these issues. This is ultimately because, regardless of which way we turn, the final lessons from the case studies about how science works will not be affected. If we were making use of the concept of a theory, there would be the issue of whether we want to say that the theory needs to include all those 'implicit premises' before it is really inconsistent. But if we are going to avoid theory-talk, then this issue doesn't arise. Of course, the question *will* arise whether we want to say that some selected set of propositions is inconsistent or not. But since no claim is being made as to *what the theory is*, no single set of propositions will be a privileged unit of analysis. We can ask: what difference would it make for the lessons about how science works if we included extra postulates to make the inconsistency 'purely logical' or if we left them out and settled for 'material inconsistency'? Suppose what makes a particular inconsistency important is that the relevant propositions were all believed to be true. If somebody has a rich conception of conceptual content, such that it is part of the conceptual content of 'Paul is married' that 'Paul has a spouse', then anybody that believes the former will believe the latter. Thus the view we take won't make any difference to the importance of the inconsistency: if what matters is that the relevant historical actors believed all of the propositions in question, we get this 'belief in all the propositions' either way. Of course, there will remain disagreement about precisely which set of propositions is 'really inconsistent', but we can ignore this question. Perhaps this is also a bad question: the best answer may simply be 'it depends how you define *inconsistent*'. The crucial point for my purposes will be that the wider lessons for how science works will be the same for both views.

This doesn't work just for 'belief', but also for most of the other issues we will care about (as indicated by questions Q1–Q6 of §2.6). There is no need for me to argue this in detail. What I will do instead for each case study is provide a natural analysis, as if I am working with a notion of 'material inference' and 'inconsistency in virtue of meaning', and I will draw relevant lessons on this basis. It so happens—I submit—that none of these lessons would change were we to add to the relevant sets of propositions further postulates to turn the 'material' inconsistency (or only 'apparent' inconsistency, depending on your point of view) into a purely logical inconsistency (or 'genuine' inconsistency, if you are so-inclined). In most cases it takes only a moment to notice that no such lessons will change. In other words, when I say that a set of propositions is inconsistent, those predisposed to generalize inconsistency to include meaning can take me literally, and those dismissive of non-logical inconsistency can interpret my claim in terms of implicit premises, and this won't make any difference to what really matters. That is, it won't make any difference to what the different cases teach us about how science works, and the lessons which might be drawn for contemporary science. Working in terms of 'pointedly grouped propositions' instead of 'theories' makes this possible: we simply won't need to ask the question whether the theory really includes relevant 'background' or 'auxiliary' assumptions or not.

2.8 Overview

This completes my presentation of the framework of this investigation. In short, in what follows I will be concerned to assess the principal (internal) inconsistency claims which have been made in the literature thus far, to consider the importance and character of the inconsistencies I find, and to discern any potential lessons for contemporary science and philosophy of science. This will be done without putting any weight on the concept *theory*: instead, my goals will be achieved through asking and answering questions such as Q1–Q6 given in §2.6, above. The guiding question will always be 'what is interesting or important about the fact that *that particular* set of propositions is inconsistent?'

Much of this work has an historical character: for example, questions Q2–Q4 are historical questions. Why so historical? Well, simply because I want to find out how science *really did* work, and the sense in which there *really were* inconsistencies in the history of science. There are more subtleties to this than are first apparent. I have said that for an inconsistency to be important for philosophy of science the set of propositions in question will have to be *historically relevant*. But this isn't a black and white issue: there is a degree to which propositions can be 'historically relevant'. Of course, one can turn to the substrate of science to look for the relevant sentences, but then there is the question of which propositions these sentences were meant to represent. Then there will be some propositions (or sentences representing propositions) bluntly stated, some which are not stated explicitly but easily inferred, others that might be contentiously inferred, and everything in-between. Then we will find propositions committed to by large communities of scientists, others committed to by a handful of individuals, others committed to tentatively by a single individual and, again, everything in-between. Some of these propositions may have been grouped together by the scientists themselves for one reason or another, others may never have been grouped together but nevertheless naturally go together for one reason or another. Others will have little to do with each other.

So the historical relevance of the propositions in question, and the character of the scientific commitment to those propositions, will both require historical work. But this isn't history of science; it is, rather, historically informed philosophy of science. Thus I will mostly make use of secondary sources. I am not suggesting that philosophers ought to do real history of science, but instead that they sometimes ought to just do *better* history of science than the status quo. The appropriate level of historical awareness depends on the investigation at hand. For my purposes, secondary sources are detailed and reliable enough, for the most part.[11]

Most of the serious allegations of internal inconsistency thus-far made within the philosophy of science literature will be considered. I start in Chapter 3 with one of the longest standing and most often cited inconsistency claims: that of 'Bohr's theory of the atom'. I will argue that any of the normal allegations do not carry much philosophical import, but that an important inconsistency emerged with Ehrenfest around 1917 which was finally noted by Pauli in 1926. Then, in Chapter 4, I turn to the most recent inconsistency claim, that of classical electrodynamics, courtesy of Mathias Frisch. He first published the claim in 2004, although it was inspired by Shapere (1969) and ultimately derives from remarks in Landau and Lifshitz (1951). Here I will argue that the inconsistency is more significant for philosophers of science than the early critics (Muller 2007; Belot 2007) have appreciated, but that it isn't as important as Frisch himself originally supposed. But certainly it does teach us interesting things about the delineation of domains of phenomena in science, and the motivation for collecting certain propositions together (as 'pointedly grouped propositions'). In Chapter 5, I turn to Newtonian cosmology, which has enjoyed a long history of being labelled 'inconsistent'. The claim was first made by an astronomer (Seeliger) in 1895 and has been fiercely debated from time to time, by both scientists and philosophers. The historical relevance of the relevant propositions is especially revealing in this case, and the pitfalls of using the word 'theory' and theory-names are especially vivid. In addition there are important lessons to be learnt about the concept of a 'truth-preserving inference' in mathematics. This latter point leads us to Chapter 6 and my final major case study: the early calculus of Newton

---

[11] Occasionally in this book I refer to the 'real' history of science, contrasting this with the a-historical reconstructions one sometimes finds in the philosophical literature. In fact there is widespread agreement that there is no such thing as 'real' or 'pure' history of science: history of science requires a perspective, and so any account is inevitably 'theory laden' (see Schickore 2011 and the references therein). But actually the theory-ladenness of history is a matter of degree: what I mean by the 'real' history of science is simply history of science at the 'pure' end of the reconstruction spectrum.

and Leibniz. This is often listed as an example of an important 'inconsistency in science' although the subject matter is of course mathematical. Thus there are rather different lessons to learn, although there are some interesting and important relationships with some of the other cases. Chapter 7 briefly surveys four more inconsistency claims which have been made, of Aristotle's theory of motion, Olbers' paradox, the classical theory of electrons, and Kirchhoff's theory of the diffraction of light at an aperture. The inconsistency in Kirchhoff's theory is one of the most obvious, but as we will see this doesn't necessarily mean that it is the most serious or important.

These eight examples provide us with a significant cross-section of (allegedly) inconsistent science, and the lessons learnt along the way are articulated and discussed further in Chapter 8, the conclusion. Perhaps the most important conclusions are really *meta*philosophical, in the sense that I will be criticizing the very framework of relevant philosophical debates (especially those already mentioned in Chapter 1). For example, despite a 'received view' in philosophy of science that there are many inconsistent theories in the history of science, I will conclude that many of the widely accepted 'inconsistent theories' are not really inconsistent in any important or interesting sense after all. Of course, given my method, I need to be aware of two obvious objections when I fail to find the inconsistency of a set of propositions to be interesting/important:

1. Perhaps there is a sense in which the inconsistency of those propositions is interesting/important that I have overlooked.
2. Perhaps there is another set of (closely related) propositions, the inconsistency of which is truly interesting/important. (I've focused on the wrong set of propositions.)

In the case studies I am highly conscious of these possibilities, and yet I still find that classic cases of 'inconsistency in science' do not deserve their status.

This problem arises (I will argue) because of a continuing disconnect between philosophy of science and (i) history of science, and (ii) science itself. As I have already noted, theory eliminativism helps here, because in eliminating *theory* one is forced to engage with science and/or the history of science in order to explain why one's point is an interesting or important one. Further, if one eliminates *theory* one also subdues a major tendency in philosophy of science to assume that one can say very general, substantial things about 'how science works'. As we progress I will argue that many important differences between the examples of 'inconsistent theories' have been overlooked, in part because their idiosyncrasies have been lost behind the common word 'theory'. Most of the literature on inconsistency in science makes certain uniformity assumptions about the scientific record, such as that there are 'many inconsistent theories in the history of science' and goes from there, seeking for example to explain how reasoning continues 'in such cases', and what general lessons we can learn about scientific goals, rationality, etc. The 'theory eliminativist' method outlined above ensures that the differences between the cases are brought to light, and that the common 'craving for generality' (Pincock 2010: 121) is subdued. Due to the individuality of each case, a general 'theory of inconsistency in science' is not to be expected.

Nickles has written 'not all inconsistencies are equal' (2002: 20). In what follows I show that it might be more accurate to say 'all inconsistencies are unequal'. If true, many of the debates about how we do/should reason with 'inconsistent scientific theories', and the lessons we can learn from past inconsistencies for contemporary inconsistencies, can be seen as based on a false premise. Science is less uniform than many philosophical projects assume, just as Kuhn already argued 50 years ago. Thus it would seem that the lessons of the 'historical turn' in the philosophy of science were not taken seriously enough: some philosophy is still 'empty' because due attention is not paid to history of science (as Lakatos warned). To adapt Kuhn's famous words:

History, if viewed as a repository for more than anecdote or chronology, could produce a decisive transformation in the image of *inconsistent* science by which we are now possessed.

But whereas—for many—Kuhn used history to show science to be less rational than was assumed (in a certain sense), here history will show science to be *more* rational than is assumed, in the sense that inconsistencies in science are fewer and less serious than a casual reader of the philosophical literature would ever believe.

But of course, so far this is mostly bald statement; the proof will be in the pudding.

3
Bohr's Theory of the Atom

3.1 Introduction

As is well-known, in 1913 Niels Bohr published a trilogy of articles arguing for a new conception of the structure and internal mechanism of atoms, especially insofar as atoms interact with electromagnetic (EM) radiation (Bohr 1913a). It was immediately clear that the proposal had some important virtues. For the first time a theory was presented which was able to explain in detail the pattern of spectral lines which had long been associated with hydrogen. That is, the theory was able to explain why hydrogen emits and absorbs light at only certain specific frequencies. But better than this, in a short period of time the theory succeeded in not only explaining the phenomena it was, in some sense, designed to explain, but in making successful predictions and explaining new phenomena. Almost immediately Fowler (1913) challenged Bohr to explain the spectral lines of (what came to be known as) ionized helium, and suggested that this would be a point at which the theory fell down. But the reply (Bohr 1913b) showed that, if one was more charitable than Fowler about the content of the theory, the spectral lines of ionized helium could be predicted exactly. When Bohr constructed the theory he had no explicit thought of explaining the spectral lines of ionized helium; accordingly, this latter achievement really caught the attention of the scientific community. On hearing of it Einstein is reported as stating, 'The theory of Bohr must then be right' (cited in Pais 1991: 154). As Pais writes,

Up to that time no one had ever produced anything like it in the realm of spectroscopy, agreement between theory and experiment to five significant figures. (Pais 1991: 149)

It is widely acknowledged that this is precisely the kind of thing which both *does* and *should* persuade scientists to take a theory seriously.[1]

However, at the same time the scientific community was pulled in quite the opposite direction. At the end of the nineteenth century Lord Kelvin spoke for many when he claimed that all that remained for physics was a certain amount of 'mopping up' work, with all the big ideas already in place. Certain developments had started to shake this confidence between 1900 and 1913—such as the success of Planck's constant and Einstein's 'photon'—but nothing was to prepare the community for the departure from nineteenth-century physics demanded by Bohr's famous postulates of 1913. As Kramers and Holst put it in 1923, '[T]he impossibility of retaining [electrodynamics] in its classical form was presented in a much clearer way than ever before' (Kramers and Holst 1923: 117).

Putting Bohr's theory in its historical context, then, claims of *external* inconsistency are to be expected. That is, Bohr's theory conflicted with many dearly held and longstanding beliefs, and for many this in itself was a great downfall. Many criticisms of the day can be interpreted as referring to *external* conflicts:

This is nonsense! Maxwell's equations are valid under all circumstances, an electron in an orbit must radiate. (von Laue 1914, cited in Jammer 1966: 86)

[The theory is] in greater or less contradiction with ordinary mechanics and electrodynamics. (Schott 1918: 243)

But the external inconsistency of Bohr's theory is certainly *not* the focus for the vast majority of the relevant literature. What allegedly makes Bohr's theory so philosophically important is that it was so successful despite being *internally* inconsistent or *self*-contradictory.

Of course whether one speaks of a given conflict as internal or external depends on how we characterize the theory, on what we take the content of the theory to be. How should this be decided? This is where all of the difficulties noted in the previous chapter come in. Bohr's theory might

---

[1] I discuss this in some detail in Vickers (2012). The theory also accounted qualitatively for the size of an atom and various empirical laws such as Whiddington's law and Bragg's law (Hettema 1995: 312ff.). It also continued to produce results as years progressed. In 1916 Epstein (upon explaining the Stark effect) wrote, 'We believe that the reported results prove the correctness of Bohr's atomic model.... It seems that the potentialities of quantum theory... are almost miraculous' (Jammer 1966: 108). See also Kragh (2012).

seem on the face of it to be one of the most obvious examples of a theory with unambiguous content upon which we can agree. But on closer inspection the question of the 'content of the theory' is far from straightforward, even in this case. Paradoxically enough, most presentations of the theory depart from the original 1913 papers, so that the term 'Bohr's theory' ends up referring to something which differs (to one extent or another) from the theory Bohr himself actually committed to. Bohr originally presented five postulates (Bohr 1913a: 874–5), but the fifth was soon dropped and in subsequent publications—by Bohr and others—the content of the other four was not held constant (especially what I will shortly present as 'P2'). Another issue concerning the 'content' of the theory—and crucial to the question of inconsistency—is the relationship between Bohr's postulates and 'classical electrodynamics', as we will see.

Such issues notwithstanding, ever since Jammer (1966) and Lakatos's (1970a) seminal paper of 1970 which describes it as 'a research programme progressing on inconsistent foundations', Bohr's theory has been widely cited as the example par excellence of an internally inconsistent theory. For example, da Costa and French speak of 'two contradictory propositions within... Bohr's theory of the atom' (1990: 186). But, when it comes to characterizing the way in which the theory is inconsistent, important disagreements arise. And there are also some—Bartelborth (1989), Hendry (1993), and Hettema (1995)—who argue that the theory is not internally inconsistent at all.[2] In addition at least some of those working at the time thought the theory to be consistent. Even as late as 1923 Rutherford was prepared to write,

For the first time, we have been given a consistent theory to explain the arrangement and motion of the electrons in the outer atom. (In Kramers and Holst 1923: xi)

The received view in the philosophical literature entails that Rutherford was just wrong about this.

At the heart of these disagreements is the concept *theory*. As is well-known, what is sometimes called 'old quantum mechanics' was a particularly messy scientific period, a period in constant flux. However, so entrenched is the concept *scientific theory* that philosophers have persisted in using it even

---

[2] In a forthcoming paper Michel Ghins also makes this claim.

in this context. The fact that the term 'Bohr's theory of the atom' is so regularly used testifies to this: the term immediately suggests a single 'thing' the identity of which (or at least *most* of the identity of which) we can all agree upon. But despite the use of this single noun term, there are various different ways in which the term is intended in the literature. Little wonder, then, that some find 'the theory' inconsistent in one way, others in another way, and some not at all.

These difficulties are surmounted by the eliminativist method to be applied here. Whether the inconsistencies discussed in the literature are interesting or important inconsistencies can be decided simply by identifying the relevant propositions which really are inconsistent, and then assessing the historical relevance and scientific character of those propositions. Three different inconsistency claims stand out in the literature, and will be considered in turn in §3.2. In §3.3, I turn to two further inconsistencies, one courtesy of Smith (1988) and another which so far has not been given much philosophical attention. In §3.4, the Conclusion, I consider afresh what sense can be made of the claim—so widely accepted in the literature—'Bohr's theory of the atom was inconsistent'.

3.2 Three Inconsistency Claims

Reconstructions of Bohr's theory have been in steady supply since 1913. Common to virtually all of them are (paraphrases of) at least some of Bohr's original postulates, as summarized in the final pages of his 1913 trilogy (Bohr 1913a: 874–75). Here is a typical reconstruction (an informal reconstruction will be sufficient, since it will turn out that no inconsistencies will depend on the fine details of these postulates):

(P1) Electrons orbit the nucleus of an atom analogously to planets orbiting the sun, swapping the gravitational attraction for a Coulomb electrostatic attraction.

(P2) Only certain orbits are possible: for hydrogen the only possible orbits are those for which the energy of the electron takes a value $E_n = \frac{hR}{n^2}$, for some integer n and with R the Rydberg constant (or, equivalently, for which the angular momentum of the electron is an integer multiple of $h/2\pi$).

(P3) Radiation is only emitted/absorbed when an electron 'jumps' from one possible orbit to another—this is a 'quantum transition'. The relation between the energies of the two electron orbits and the frequency ν of the radiation emitted/absorbed is $\Delta E = h\nu$.

No claim has ever been made that (P1)–(P3) are inconsistent. So what do we need to add to these postulates to reach inconsistency? And on what grounds do we add these extra propositions and declare the result 'an internally inconsistent theory'? Many articles in philosophy of science simply state the 'fact' that Bohr's theory of the atom is internally inconsistent and go on to make certain general claims about the character of theories, scientific rationality, or whatever. Those papers which do give a story of how the alleged inconsistency manifests itself usually don't give much detail, and one never finds a rigorous derivation of a contradiction from given assumptions.[3]

However, within the literature one can distinguish three different features which are usually identified as the problem areas:

(1) The discreteness of the energy states
(2) The 'quantum transitions'
(3) The non-emission of radiation from a charged, orbiting particle

These will be considered in turn, in §§3.2.1, 3.2.2, and 3.2.3.

3.2.1 *Discrete energy states*

Brown (1992) identifies the inconsistency of the theory as a conflict between quantum and classical principles. He writes,

Bohr's approach provided limited classical descriptions of the stationary states, but no account of transitions between them. . . . This combination of classical and non-classical principles was a logically risky game. . . . The principles are inconsistent with each other. (Brown 1992: 399)

And da Costa and French (2003) pick upon this passage, writing,

Brown emphasises the point that classical mechanics was taken to apply to the dynamics of the electron in its stationary state, while quantum theory was brought into play when the transition between discrete states was considered—this discreteness contradicting classical physics, of course. (da Costa and French 2003: 91)

---

[3] Hendry (1993: ch.3) raises this concern: 'There is . . . no trace of a contemporary "logician's proof" of the inconsistency of Bohr's atomic model.'

Now, there are actually two separate claims in these quotations. Brown identifies the *transitions* of electrons as 'quantum' and as conflicting with the classical, whereas da Costa and French state that it is the 'discreteness of the energy states' where the conflict with classical physics arises. I'll consider the latter suggestion first, in this section. How could the discreteness of the energy states lead us to inconsistency?

Following the eliminativist method, the first step is to consider precisely which propositions make up the inconsistent set. Well, postulate (P2) will certainly feature, since this gives us the discreteness of electron energy states. The big question is just what classical physics we would need to introduce to reach the contrary conclusion, that electron energy states *cannot* be discrete. Is there a principle to the effect that allowed energies of physical systems must be continuous in all contexts?

Sometimes one hears claims to the effect that 'classical physics is inherently continuous'. As they stand such claims are too vague to be true or false. Certainly there is nothing 'non-classical' about finding the electric charge of oil drops to always be some multiple of a fundamental, discrete unit of charge. But in the present context we are talking about discreteness of the energy of electrons, which might seem a bit different. Marchildon (2002), for example, writes,

In classical physics, a system's energy is a real variable and its allowed values fill a continuum. (Marchildon 2002: 4)

If this is meant to stand for *all possible contexts*, then we have our contradiction with postulate (P2).

However, it isn't clear that Marchildon's statement can remain true (true to scientific history) if it is meant to apply to all possible contexts. There are certain, purely classical (on any reasonable definition of 'classical'), mechanical systems where only certain energies of the system are stable, such that if one starts the system at some other energy level it will quickly collapse into one of the 'allowed' energy states. A familiar example of this sort of phenomenon is the well-known fact that an object—for example a violin string or a cantilever—experiences only certain, discrete modes of vibration. And each such mode has an associated energy. It might be objected that such objects *can* have any real-valued energy, it's just that the energies associated with the modes of vibration are in a sense discrete. But then, so too in the case of Bohr's postulate (P2), there is no claim that electrons *cannot* have any real-valued energy, but instead a claim that electrons can only have the specified, discrete energies if they are to orbit the nucleus of an atom.

The question I am asking is whether a 'continuity principle' can be stated which does justice to classical systems, but which forbids Bohr's model. Such a principle must be permissive enough to allow for discreteness in *some* sense, since classical physics is no stranger to certain types of discreteness. But it must be prohibitive enough to mean that Bohr's postulates—and in particular (P2)—are contradicted. Whether this *can* be done is not the issue. Suppose we did consider various formulations of a 'continuity principle' for classical physics, and found one that fit the bill. Would we then have an inconsistency that was interesting or important, that tells us something about how science works? Not unless the principle, in that particular form, played some serious role in the relevant history of science. Otherwise, in what sense would there have been an 'inconsistency in science' here at all, as opposed to an inconsistency only in our present-day, a-historical, philosophical reconstruction of that science?

I submit that no such principle will be found explicitly stated in the relevant substrate of science. One possibility would be to look to the practice of pre-quantum scientists and argue that such a principle was *guiding* science, even if it wasn't explicitly articulated. There is no need for me to follow this possibility up: the burden of proof lies squarely with those who wish to claim that there is an historically relevant inconsistency here. But further, *even if* this project could be successful, there is an important sense in which the inconsistency would still fail to be important or interesting. This comes to the fore if we turn our attention from the historical relevance of the propositions to the question of in what sense they could be 'pointedly grouped'.

Let's suppose for the sake of argument that the 'continuity principle' in question is indeed historically relevant in the sense that scientists working in the 'classical physics' tradition committed to it explicitly or implicitly. It certainly doesn't follow that those committed to Bohr's postulates had inconsistent commitments, since they were obviously rejecting that principle when they committed to Bohr's postulates. This is obvious, because it's so clear that Bohr's postulate (P2) introduces a certain discreteness, or discontinuity. In the process of making an explicit commitment to this postulate, any prior commitment to some 'continuity principle' is being dropped, or at least suspended.

The objection might be made that Bohr's theory also *makes use* of this principle. Or at least that there is *some* good reason to include the principle in 'the content of Bohr's theory'. If it was going to be an 'internal' inconsistency, such a claim would have to be made. But I for one cannot see any way in which one might argue that Bohr's theory implicitly employs, or otherwise demands commitment to, the kind of continuity principle I have been discussing. At least if the principle is going to have the kind of prohibitive strength it would need to have to contradict postulate (P2).

This gives some sense of the reasons one might have for steering clear of labelling Bohr's theory inconsistent for *this* reason (the discreteness of orbits). For one thing, at least some of the propositions needed to reach genuine inconsistency are not historically relevant, at least in any obvious way. But in addition there doesn't appear to be a sense in which the relevant propositions are 'pointedly grouped', except for the fact that they are inconsistent. For example, they don't appear to be doing any real scientific work *as a group*, and they don't feature in the commitments of any *individual* scientist at a single point in time. But anyway, it isn't clear that this inconsistency claim is the most serious candidate for the inconsistency 'of Bohr's theory'. It's time to turn to another possibility.

3.2.2 *Quantum transitions*

As noted in the previous section, Brown (1992) focuses on the *transitions* between energy states (as opposed to the discreteness of the states themselves) as the source of the inconsistency of the theory. But precisely which set of propositions is the relevant inconsistent set here? And how is the inconsistency of that set supposed to provide us with an insight into how science works, or otherwise be an interesting or important inconsistency?

The idea appears to be that the quantum transitions are really quantum 'jumps', with a break in the continuity of the relevant electron's worldline. Bohr's postulates tell us that electrons do not have worldline continuity, but classical physics insists upon worldline continuity. Thus we reach inconsistency.

Again, historical relevance here is doubtful. In Bohr's original 1913 trilogy there is no mention at all of 'jumps', but instead of 'transitions' between states. Taken on its own, Bohr's original trilogy gives one the impression of electrons traversing the allowed orbits on continuous worldlines. Indeed, Bohr had no reason at this point to make the more radical claim that electrons have discontinuous worldlines, and he wasn't read as making this claim either. Upon reading the trilogy Rutherford wrote to Bohr in 1913, 'It seems to me that you have to assume that the electron knows beforehand where it is going to stop' (Pais 1991: 153). Clearly the electron is conceived to travel from one orbit to another, without any break in worldline continuity.

But even if there was discontinuity it still isn't obvious that we have good reason to put the relevant propositions together in a group. For example, were the propositions ever jointly committed to in some sense? Surely not: any scientist making a serious scientific commitment to quantum *jumps* in Bohr's theory was obviously giving up any previous commitment to the ubiquitous worldline continuity of particles. The inconsistency of Bohr's theory is supposed to be something special, but if *this* were the inconsistency in question it would certainly not be special. Instead, it would be simply a case of scientists giving up on the idea that a certain principle they previously adhered to is ubiquitously applicable. The inconsistency is then merely a temporal one, and shows only what everybody already knows: that scientists sometimes change their minds about things.

There is another sense in which the relevant propositions might be 'pointedly grouped', such that the inconsistency ends up being interesting and/or important (and which will be important throughout this book). This is that the propositions, although not really 'committed' to by scientists in an obvious way (e.g. doxastically), are nevertheless *used* together for certain purposes. This might happen, for example, because although scientists sometimes believe a proposition to be false, it is nevertheless useful for the purposes of scientific inference, for example because it is a simplified idealization of a more complicated, intractable truth. So was the relevant proposition about continuous worldlines *used* alongside Bohr's postulates?

Once again, historical relevance seems to speak against this possibility. Suppose we notice that Bohr is assuming worldline continuity for electrons *in orbits*. We might then claim that Bohr *infers* worldline continuity for electron trajectories in orbits *from* the broader principle of 'ubiquitous worldline continuity of particles'. Then Bohr would be *using* (even if he wasn't doxastically committed to) inconsistent propositions. But why reconstruct the history in this way? The only reason to do this, it would seem, is if one is keen to show that 'Bohr's theory is inconsistent'. In other words, it is an example of deciding what one wants to show, and then reconstructing the history in such a way that one can show it. The obvious alternative is to see Bohr as using only what he needs, and only what he explicitly *does* use: that worldline trajectories of electrons in orbits are continuous—leaving open the possibility of worldline discontinuity elsewhere.

In summary, then, the set of propositions we need to put together to make this particular inconsistency claim stick are not historically relevant *as a set*. For example, there is no place in the substrate of science where one will find the propositions stated side by side, or all assumed in a single piece of reasoning. Certainly some would have signed up to worldline-continuity even after 1913, but these would have been individuals who rejected an interpretation of Bohr's theory which called on discontinuous 'jumps'. Thus there was no need in this case for any individual to make inferential restrictions to avoid logical explosion. We do not need,

[A]n appropriate closure relation on the set of principles accepted by Bohr to replace the trivial closure relation we get from classical logic. (Brown 1992: 405)

That is, unless Brown is talking about different principles to those considered in this section. One possibility for this will arise in §3.2.3.2.

3.2.3 *Non-emission of radiation*

The two inconsistencies discussed in the previous sections are sometimes mentioned in the literature, but they are not, perhaps, especially serious, or especially common, inconsistency claims. The more common inconsistency claim involves the non-emission of radiation from a charged, orbiting particle. The orbiting electrons are accelerated charged particles, and according to Maxwell–Lorentz 'classical' electrodynamics (CED) they are required to emit a constant stream of radiation, as opposed to the intermittent quanta of Bohr's theory. Most seriously there should be no state—the 'ground state'—where the amount of energy emitted is a maximum. On the classical picture, the electrons should continue to emit radiation, causing a loss of energy and a consequent suicidal spiral trajectory into the nucleus. Several authors highlight this aspect of the theory as the focus of an internal inconsistency. From the quotation given above in §3.2.1 da Costa and French (2003) continue,

> However it is not only in the discreteness of the states that we have conflict between quantum and classical physics but also in . . . the assertion that the ground state was stable, so that an electron in such a state would not radiate energy and spiral into the nucleus as determined by classical physics. *This* is the central inconsistency. (da Costa and French 2003: 90)

In his 1970 (1970a) paper Lakatos's discussion, although ambiguous in places, can plausibly be taken as making this same point.[4]

Now as already indicated this is sometimes taken to be the source of an *external* inconsistency: the classical physics which insists upon constant emission of radiation from a charged, accelerating particle can be conceived as external to Bohr's theory such that Bohr's theory is not itself inconsistent, but rather inconsistent *with* something else. Priest (2002) is one author who refers to this explicitly as an *internal* inconsistency, writing, 'Bohr's theory . . . included both classical electrodynamic principles and quantum principles that were quite inconsistent with them' (2002: 122f.). Shapere (1977: 561) on the other hand refers to such classical principles as 'background information'. But as already noted, the question of whether the inconsistency is internal or external is dissolved by the method advocated here. The only question is whether the propositions which *are* inconsistent count as historically relevant, and reveal a significant inconsistency *in some sense*. In particular the question is whether there is a good reason, historical or scientific, why the propositions required to reach contradiction belong together, as pointedly grouped propositions. Two of the most obvious ways in which this *might* be the case are considered in the next two sections.

3.2.3.1 *Doxastic commitment* One obvious reason to put Bohr's postulates and classical physics together as pointedly grouped propositions would be if they all play a role in Bohr's explanation of atomic spectra. Clearly Bohr's postulates P1–P3 do play a role, but what of 'classical physics'? Well, Bohr did draw on classical physics to characterize the electron orbits: electrons are assigned an angular momentum, and said to follow a smooth, periodic

---

[4] This is how Bartelborth (1989: 221) understands Lakatos, and it is this aspect which he also focuses on.

trajectory; they are taken to be charged particles which are held in their orbits by a Coulomb attraction to the nucleus. So Bohr needs *some* of CED to characterize the orbits. Does this mean that he is thereby committed to *all* of it?

Bartelborth (1989) argues not.[5] He writes,

[T]he only necessary theory-element from classical electrodynamics for Bohr's theory is quasi-electrostatics for point particles, because what Bohr really needed from classical electrodynamics was the concept of electric charge and Coulomb's law. (Bartelborth 1989: 221)

In other words, if we consider classical electrodynamics as a collection of objects (equations, models, Gauss's law, the Lorentz force equation, etc.) then one can simply pick out an independent subset of those objects—which may be called 'quasi-electrostatics'—which are all that Bohr needs to achieve what he wants to achieve.

The point is not whether this *can* be done, but whether it is relevant to the thinking of scientists at the time, or perhaps whether it *ought* to have been relevant to the thinking of scientists at the time (in some sense). What is clear is that Bohr and those following him rejected CED as universally applicable. In his third and most persuasive explanation of the spectral lines of hydrogen and ionised helium, Bohr drew on an early version of the 'correspondence principle'.[6] The idea was that CED can be applied to electrons with increasing confidence as the principal quantum number n increases. Or, in the case of the hydrogen atom, as the electron orbiting the nucleus gets further and further from that nucleus, its behaviour becomes more and more classical. The basic idea here is simple: Bohr assumes that CED can be trusted, more or less, in its traditional domain of application. But the obvious flipside of this is that Bohr is assuming that CED does *not* apply to electrons strongly bound to atoms (with relatively low quantum number n). As he puts it in an article from 1921, he is looking 'to obtain a rational generalization of the electromagnetic theory of radiation' (Bohr 1921: 104).

---

[5] Hendry puts forward essentially the same argument, independently of Bartelborth, in chapter 3 of his PhD thesis (1993); Hettema (1995) agrees with Bartelborth but doesn't provide an independent argument.

[6] For the details of this derivation, including an explanation of why it is Bohr's 'most persuasive' explanation of various spectral lines, see Vickers (2012).

Of course, how interesting or important the inconsistency is doesn't hang only on what Bohr thought. As I've already noted, what many people think of—and thought of at the time—as 'Bohr's theory of the atom' is something slightly different to what Bohr wrote down in the original 1913 trilogy. But it's easy enough to see that those following Bohr also rejected the universality of CED. The attitude of the community is well represented by Millikan in his presentation of Bohr's theory in 1917:

Bohr's first assumption... when mathematically stated takes the form: $\frac{eE}{a^2} = (2\pi n)^2 ma$, in which e is the charge of the electron, E that of the nucleus, a the radius of the orbit, n the orbital frequency, and m the mass of the electron. This is merely the assumption that the electron rotates in a circular orbit.... The radical element in it is that *it permits the negative electron to maintain this orbit* or to persist in this so-called '*stationary state*' without radiating energy even though this appears to conflict with ordinary electromagnetic theory. (Millikan 1917: 207f., first emphasis added)

The point is clearly made: the theory doesn't include the parts of electrodynamics which give rise to the self-radiation of accelerated charged particles. As Jeans wrote in 1924, 'The complete system of dynamics, of which it [the quantum theory] is a part, has not yet been found' (1924: 36).[7]

Of course, this still leaves a genuine inconsistency in the history of science here: Bohr's postulates are inconsistent with CED, and in particular with the MEs. But the inconsistency of Bohr's theory—as the philosophical literature presents it—was always supposed to be something special. All we appear to learn from the inconsistency of Bohr's postulates with the MEs is that sometimes well established scientific laws and principles are left behind when they fail to do justice to new phenomena—but this is hardly news, hardly an interesting lesson about how science works.

One option for rejoinder is that Bohr cannot help himself to Coulomb's law without committing to the Maxwell equation from which it is derived: Gauss's law. And then one might claim that, given the interrelationships between Gauss's law and the other MEs which integrate them into a coherent whole, Bohr must also be committed to the whole of CED. Turning these bare claims into persuasive arguments strikes me as

---

[7] Some modern reconstructions do justice to this rejection of CED by Bohr and his followers, explicitly presenting the theory as making a commitment to only electro*statics* (and not electrodynamics). See e.g. Norton (2000: 80). Cf. Hendry (1993).

an ambitious project. Bartelborth (1989) claims that the fact that CED is divisible and has independent parts 'is proven by many structuralist reconstructions of physical theories' (1989: 222). But one does not need to commit oneself to Bartelborth's particular brand of structuralism in order to rebut the rejoinder. Instead one can shift the burden of proof, and simply ask for an argument as to why Bohr and others would *have* to be committed to the rest of CED if they were committed to Coulomb's law, or Gauss's law.

I should clarify that in the previous section by 'makes a commitment' I have meant 'believes' or 'takes to be true' or 'takes to be a (serious) candidate for the truth', or similar. In other words this section has all been about what I am calling *doxastic* commitment. But there is another sense of 'commitment': Bohr and his followers obviously *were* committed to CED in an important sense, by virtue of the fact that they sometimes *used* CED. This leads to a new way in which the propositions in question might be 'pointedly grouped'.

3.2.3.2 *Inferential confidence* In §2.8, I mentioned a possible objection that might arise if I claim that an inconsistency is not interesting or important: it could just be that there is a sense in which it is interesting or important that I have overlooked. If the previous subsection were all I had to say on the 'non-emission of radiation' inconsistency, then that would be the case here. The objector would say that I am being uncharitable and arguing against a straw man, that I am missing the sense in which the inconsistency in question is supposed to be interesting or important, or at least eye-catching when one looks to the history of science.

In his 1990 paper Brown writes as follows:

[T]he radiation emitted by the atom is assumed to be describable in terms of classical electrodynamics (CED), while the emission and absorption processes, as well as the behaviour of electrons in stationary states, are accounted for in terms manifestly incompatible with CED. (Brown 1990: 285)

Here we have the same conflict between Bohr's postulates and CED, but Brown provides a new reason why we might think of the relevant propositions as 'pointedly grouped'.[8] It might well be obvious (the claim goes) that

---

[8] That this is Brown's preferred focus of inconsistency is signalled by the fact that he reiterates *this* aspect (but not the one noted in the previous section) in his 2002 paper (90).

Bohr and his followers are not committed in a doxastic sense to the conflicting propositions, but they nevertheless *made use of* inconsistent propositions in the same scientific context. The relevant community were *reasoning with* the relevant set of inconsistent propositions, even if they didn't believe all of them, or even if they believed some of them to be definitely false.

Now this seems to be a fair claim in a sense. One can find the relevant scientific community making use of CED *and* Bohr's postulates. Of course, even if they rejected CED as universally true, they could hardly be expected to stop using it in various contexts given how successful it was. And, as Brown points out, they used it hand in hand with Bohr's postulates, for example when they first considered the behaviour of electrons in atoms, and then considered the behaviour of the radiation emitted by those atoms. Isn't this use of conflicting principles in some sense alarming, or surprising?

One option might be to reject the claim that we have an inconsistency here at all. It *looks* like we have inconsistency, but only because scientists are not writing down exactly what they mean. As Muller has recently put it in a different context (which will be the subject of the next chapter),

[P]hysicists are notoriously sloppy in this respect: a majority of the exact equality signs (=) in most physics papers, articles, and books mean approximate equality ($\approx$). (Muller 2007: 261)

Accordingly we might claim that Bohr and his followers may have *used* '=' when they drew on various equations of CED, but that, given their commitment to ultimately finding a new, better dynamics, they really meant '$\approx$'. In this way, one can finesse away the alleged inconsistency. We no more have inconsistency than if somebody colloquially states 'I can't believe he didn't come' whilst at the same time fully believing he didn't come. We have literal inconsistency, but no inconsistency in intended meaning.

But this does not answer the point at issue here. We are accepting that the relevant scientific community rejected CED in *some* sense, such that, if asked, they would have only committed to its *approximate* validity. But nevertheless, the point remains that they *reasoned* with the relevant equations *as if* they were true. This is a sensible way to reason with approximately true equations, since applying truth-preserving inferences to approximately true equations gives one a chance of staying close to 'the truth'. And how else would one proceed with approximately true equations, in particular where (as in this case) one doesn't know the sense in which the equality is approximate? But then at the level of *reasoning* (if not belief) the question of inconsistent commitments looms large: how were scientists able to reason with, or thought it rational to reason with, mutually inconsistent propositions? We saw in Chapter 1 how it is possible to infer absolutely anything with inconsistent propositions. How then can one have confidence in anything that is inferred from an inconsistent set?

Actually, this move from it being *possible* to infer anything from an inconsistent set to having no confidence in anything inferred from such a set is much too quick. Consider the set of propositions Q_G (introduced in Chapter 1) consisting of some extremely basic truisms of logic and mathematics—Robinson Arithmetic—and Goldbach's conjecture, that any even number is the sum of two primes. This could in fact be inconsistent, because we don't know whether Goldbach's conjecture is true; let's suppose for now that Goldbach's conjecture is false so that the set of claims in question, Q_G, *is* inconsistent. In other words there is at least one even number that *isn't* the sum of two primes. What happens when we reason with this inconsistent set?

Clearly any (truth-preserving) inferences made with Robinson Arithmetic alone will lead only to mathematical truths. Any problems will only come in if we include Goldbach's conjecture. But then, we do know that many, many even numbers are indeed the sum of two primes, so that any reasoning which starts with Goldbach's conjecture but makes a universal instantiation may well lead to mathematical truth. In other words, it's perfectly possible to start with this inconsistent set and, even making use of *all* of the propositions in question, nevertheless reach true conclusions. What would it take to infer any arbitrary proposition? As discussed in §1.2.1, the only way is to first infer a contradiction or explicit contradictories (that's the point of the 'C' in 'ECQ'!). The difficulty of doing this with Q_G (assuming for the moment that Goldbach's conjecture is false) is really quite incredible. Thus, in *this* case at least, any claim that one should be wary of one's inferences because of ECQ is just a mistake.

Of course, one may reach all sorts of *false* conclusions by reasoning with the propositions in question. If Goldbach's conjecture is false, then any number of universal instantiations from it may lead to a false conclusion. But there is no special problem of inconsistency here: instead this is now about the care that may sometimes be required when one knows that at least one of the assumptions one is reasoning with is false. But this is a ubiquitous phenomenon in science, with an associated large literature. If this is the most interesting aspect of the 'inconsistency of Bohr's theory' then once again we find that Bohr's theory is not nearly as special as has often been made out. It is, instead, an example of a quite common feature of science. Scientists regularly reason with assumptions at least one of which they know or believe to be false, either because the (assumed) truth is computationally intractable, or because the truth is unknown. There is usually nothing irrational about this; as I've just demonstrated, perfectly sensible conclusions can sometimes come from false assumptions. In addition, reasoning with approximately true propositions can often lead to approximately true conclusions.[9]

So much for the general story; how does reasoning from inconsistency play out in the particular case of 'Bohr's postulates + CED'? As in the Goldbach case, for ECQ to come into play we need all of the inconsistent propositions to be employed in the same piece of reasoning, and employed in a very specific way. Brown's example doesn't appear to fit the bill. For one thing, he certainly doesn't demonstrate that all of the relevant inconsistent propositions are employed in the same piece of reasoning. But even if this is the case, it would fall far short of showing that the propositions are employed in a way which leads, or even could lead, to contradictories. Again, it's hard to imagine just how explicit contradictories *could* be derived from the propositions in question unless (i) one actually sets out to derive them, or (ii) one plugs the propositions into a computer and leaves it to blindly make inferences. Needless to say, these two options do not represent how science works! What remains is the familiar problem that inferences are being made from propositions at least some of which are thought to be false. Once again, if this is all there is, then there is no special problem of inconsistency here.

But let's suppose for the sake of argument that the relevant propositions sometimes *were* used in the same context, or the same derivation, such that contradiction and ECQ threatened. There still isn't anything particularly

---

[9] A real-life scientific case of very deeply buried inconsistency is Maxwell's equations combined with Kirchhoff's assumptions about the diffraction of light at an aperture (see Saatsi and Vickers 2011: 40). Kirchhoff's theory of diffraction will be tackled in Chapter 7.

special happening here. The community in question obviously knew that they were reasoning with inconsistent propositions (this was established in the previous section). So they knew that at least one of the propositions they were reasoning with was false. In these circumstances the game changes: scientists don't blindly trust their inferences at all, but on the contrary are highly sceptical of them. Some of their inferences will be reliable, whereas some won't, but if they don't know in just what way their assumptions *are* false (as in this case), then they won't know which ones are reliable. Consider a situation where we know that $(kx^n - 31)^m = y$, and we know that $k = 50$ is approximately true (within 0.5), $x = 0.95$ is approximately true (within 0.06), $n = 10$ is approximately true (within 0.1), and $m = 5$ is approximately true (within 0.1). Carrying out truth-preserving inferences with our 'best guess' values we reach $y = -1.4$. Is this approximately true? Not if, in fact, $k = 50.5$, $x = 1.01$, $n = 10.1$, and $m = 5.1$, since then $y = 1.3 \times 10^7$. This may seem like an example tailored for dramatic results, but in real science results are often likely to be just as dramatic, given that usually many more variables and steps of calculation are involved.[10] And the point isn't just that one's error bars get bigger as one carries out more inferences, it's that sometimes we don't know which propositions in our set are approximate, or in what way they are approximate, so we can't even start to introduce error bars. Indeed, for some sorts of approximation (e.g. qualitative approximation) talk of 'error bars' will not be appropriate at all. The lesson is that if one starts with approximately true, but false, assumptions, and one employs truth-preserving inferences, one often can't be sure if those inferences will keep one close to truth, or take one far from truth. In practice, one doesn't trust one's inferences in such a case, but instead tests one's inferences empirically, or at least tries to judge on conceptual grounds whether they give largely sensible results.[11]

This gives a rough idea of how scientists reason when they know that at least one of the propositions in play is strictly false; something like this was happening in the days of OQT. But we are still at a loss as to why the inconsistency in question is interesting or important. The community knew about the inconsistency, so they knew that they were sometimes reasoning

---

[10] For real-life scientific examples where this sort of phenomenon has been explicitly noted, see Kythe and Puri (2002: 321ff.), Sitte and Egbers (2000: 324), and Mártin and Tsallis (1981).

[11] Mark Wilson has done some interesting work looking at specific cases of where our 'false' assumptions do and don't break down. See e.g. Wilson (2006) and Wilson (2013).

from falsehoods. But then the inconsistency acted just to tell them exactly that—that they were reasoning from falsehoods—and didn't play any other special role in that scientific period. The most interesting aspect of the inconsistency really seems to be that it doesn't affect the relevant science much at all, no more than the approximations and idealizations that are ubiquitous in science.[12]

At this stage I hope to have firmly laid the burden of proof at the door of those who think there is a 'special' interesting or important inconsistency to be associated with Bohr's theory. I have briefly considered the possible rejoinder that there is some *other* way in which the inconsistency of the relevant propositions is interesting or important that I am yet to consider. On this point I can't do much more than consider all the obvious ways in which the inconsistency could be interesting or important and then, having come out the other end empty handed, lay the burden of proof at the feet of those who maintain that there is such an inconsistency. This I have done. But in §2.8 I also introduced another possible rejoinder to such a conclusion: that I have focused on the wrong set of inconsistent propositions, that there is some other (closely related) set of propositions the inconsistency of which *is* interesting and important. Is that the case here?

In the case currently under discussion, which sets of inconsistent propositions could be candidates? Again, I am inclined to just shift the burden of proof here: why believe there is such an inconsistency until it has been clearly presented in the literature? But there is one new assumption I wish to introduce at this stage. I noted above that sometimes scientists work with assumptions they believe to be strictly false for one reason or another. In fact, this was also the case with Bohr's postulate P2. Postulate P2 was taken as an approximation by the community at the time; it was not stated absolutely as P1 and P3 were. Bohr made it clear that the stationary states defined by his postulates were not meant to be the only possible states. He writes, 'There may be many other stationary states corresponding to other ways of forming the system' (1913a: 22). For example, Bohr ultimately wanted to allow for *elliptical* orbits (cf. p.875). And he also notes that he has 'assumed that the velocity of the electrons is small compared with the

---

[12] It is not my goal to present an account of reasoning from assumptions at least one of which we believe to be false. The inconsistency is suitably deflated if I have shown that it is no more troublesome or interesting than approximations and idealizations in science. There is a mountain of literature concerned with the latter, to which I direct the interested reader. (E.g. Brzeziński and Nowak 1992.)

velocity of light.' The extra details would have to wait until Sommerfeld's efforts in 1916. But in the meantime, what Bohr was actually committed to in a doxastic sense was not P2, but rather the following:

(P2★) Only certain orbits are possible, and for the hydrogen atom $E_n = \frac{hR}{n^2}$ gives *at least a good approximation* to *at least some* of them.

The famous explanations and predictions can still be recovered using this axiom (although the predictions will have to be qualified as 'at least good approximations'). But now we also have a statement which the community actually signed up to in those early years of the theory, as a serious candidate for the truth (for those that were realists).

We will see this sort of move again in other case studies. Let us call it 'internalizing the approximation', since in moving from P2 to P2★ we move any approximation from the character of the commitment *to* the assumption to the content of the assumption. Note that this isn't a way to avoid inconsistency, as if that were a goal in itself. As I have already mentioned in Chapter 2, I am not trying to play down inconsistencies here: there is no agenda except to simply assess the inconsistency claims which have been made and consider the sense in which inconsistencies in science teach us interesting and important lessons. The point of internalizing the approximation is just to introduce new propositions which may bring us to consider new, interesting or important inconsistencies. And it is a strategy which we might *think* will lead to interesting and important inconsistencies, since in making the move we reach propositions committed to as serious candidates for the truth, and such a commitment to *inconsistent* propositions would be something remarkable. In the case currently at hand the strategy doesn't work, however. P2★ is a weaker proposition than P2 (P2 actually entails P2★) so if we cannot find any interesting inconsistency using P2 we will not find any with P2★.

3.3 Inconsistencies in the Later Theory

So far I have focused exclusively on the early Bohr theory. This is usually where the attention vis-à-vis inconsistency is placed. But as Bohr's theory developed, new theory elements surfaced which at least some in the community came to consider 'part of the theory', and which present the possibility of different inconsistencies to those mentioned above. In the next two subsections I consider briefly this later period. I turn first to hypotheses concerning spectral line intensities, which feature in an inconsistency claim courtesy of Smith (1988). Secondly I turn to Ehrenfest's adiabatic principle, which led to 'forbidden' predictions in 1926, and where we can find a more interesting and more important (but seldom discussed) inconsistency.

3.3.1 *Spectral line intensities*

Smith (1988) claims that 'Several parts of Bohr's theory of spectral emission were based on inconsistent proposals' (1988: 439). But the aspect he focuses on as 'most interesting' is Bohr's method of calculating spectral line intensities. Here Bohr's correspondence principle (CP) plays a role, and it is the particular role it plays that leads to inconsistency, in Smith's view. And he claims support from Lakatos (1970a) on this point. He writes,

The fact that the correspondence principle was an inherent part of Bohr's research programme was undoubtedly the reason that Lakatos referred to it as progressing on inconsistent foundations. (Smith 1988: 441)

Now on the face of it this might seem rather strange: we've already met the CP in §3.2.3—I said that it was the principle that CED can be applied to electrons with increasing confidence as the principal quantum number n increases. Stated in this modest way, one might well find it hard to believe that it could generate an inconsistency. But, in fact, it is not the CP stated *this* way that Smith is interested in. He follows Van Vleck (1926) who distinguishes between the CP and the correspondence *theorem* (Smith 1988: 440, fn.31). What I have been calling the 'CP', Smith and Van Vleck call the 'correspondence theorem'. The question then remains what exactly Smith means by the 'CP', such that it is 'an inherent part of Bohr's research programme', and such that it brings about an inconsistency.

On p.440 Smith describes the 'CP' as 'The use of classical results for intensity and relative phase data about spectral lines.' Now, the 'correspondence theorem' (what I called the 'CP' in §3.2.3) tells us that this should be allowed in the realm of large principal quantum number n, where the quantum and classical domains meet. But Bohr reasoned that if frequency and intensity can be connected in a classical way for large n, then they can also be so-connected for smaller values of n.[13]

The details of Bohr's proposal will not be especially important here. What is important is that Bohr's proposal regarding intensities was extremely controversial, and just one of several different proposals for calculating intensities made in those years. Sommerfeld made a preliminary suggestion regarding intensities in 1916, which Bohr rejected in formulating his own proposal in 1918 (Bohr 1918). Rudolf Ladenburg tried a different approach in 1921, which was also followed by contributions in the early 1920s from Born and Van Vleck, Burger and Dorgelo, and Heisenberg.[14] Bohr's doctoral student Kramers (1919) did provide some empirical support for Bohr's particular proposal, but the predictions weren't successful enough to attract widespread community support as Bohr's postulates P1–P3 did. In fact, OQT was never particularly successful when it came to spectral line intensities, and to such an extent that authors today generally don't consider 'Bohr's theory' as telling us anything at all about spectral line intensities. Eisberg and Resnick (1985) write bluntly, 'The theory...does not tell us how to calculate the intensities of spectral lines' (1985: 119) and Shapere is very clear:

[T]he Bohr theory offered no way to account for the intensities and polarizations of the spectral lines.... Use of the correspondence principle as a basis for calculating the polarizations of the lines is not considered here as a 'part of the theory.' The principle was not, in any case, very successful with regard to the intensities. (Shapere 1977: 559)

Now of course in this analysis I do not want to take a stance on whether the CP was 'a part of the theory'. The question is whether there is some point to grouping this principle with other postulates relevant to that scientific period such that any resulting inconsistency carries interesting or important lessons for our understanding of how science works. One thing to note right from the start is that we aren't going to find any large-scale community 'acceptance' or 'commitment' to the CP (as Smith interprets it), whether that be doxastic commitment or a commitment to use it for explanatory and

---

[13] See Smith (1988: 441–2), especially the quote from Bohr at the top of p.442.
[14] For further details and references see Jammer (1966: 109–18), and Mehra and Rechenberg (2000a 645–9).

predictive purposes. One does find the following sort of statement in the literature:

> The correspondence principle has... given rise to important discoveries and predictions which agree completely with the observations.... It has made possible a more consistent presentation of the whole theory, and it bids fair to remain the keystone of its future development. (Kramers and Holst 1923: 141)

But we may ask the question what Kramers and Holst mean by 'the CP' here. Although they don't provide anything like a definition, it is clear that in the above passage they have in mind a more general formulation of the principle than would be needed to calculate spectral line intensities. At the top of p.139 they say that the CP 'expresses the previously mentioned connection with the classical electrodynamics', and looking back to p.132 this 'previously mentioned connection' seems to be little more than the fact that 'the Bohr theory for large wavelengths... leads to a formal agreement with electrodynamics' (in other words what Smith refers to as the 'correspondence *theorem*').[15] Commitment to, and endorsement of, a principle allowing CED to be used to calculate spectral line intensities for small quantum number n is much harder to come by.[16]

This suggests that, *even if* the relevant set of propositions is inconsistent (a question I'll turn to in a moment), this isn't going to be a very interesting or important inconsistency. In general, as philosophers of science we care about ideas that are successful and/or take off within a scientific community in a serious way. It should be no surprise that, from time to time, certain individuals have inconsistent commitments that don't really lead anywhere and that nobody else (apart from one's own doctoral student, perhaps) is interested in. It is somewhat more interesting if somebody as famous as Bohr has such commitments, but still not especially interesting. In a serious sense we don't learn anything about how science works from such an

---

[15] In the end I think Smith does get confused by the fact that the term 'correspondence principle' is used in (at least) two different ways in the literature. He draws on a passage from Jammer (1966: 116) which refers to the 'correspondence principle', but it's not at all clear to me that Jammer has a formulation of the principle in mind that would enable one to calculate spectral line intensities for small principal quantum number n (at least without adding further assumptions). Cf. Jammer's discussion of 'Bohr's first explicit formulation of the correspondence principle' on p.110.

[16] Cf. Dirac: 'The correspondence principle always seemed to me a bit vague.... All it said was that there was some similarity between the equations of quantum theory and the equations of classical theory. I don't believe it was more definite than that' (quoted in Mehra and Rechenberg 2000b: 108). See also Darrigol (1992: 79ff).

inconsistency, because it just isn't 'science working' when a single individual has a scientifically suspect idea that doesn't go anywhere.

Is Smith just wrong, then, when he says that 'the correspondence principle was an inherent part of Bohr's research programme'? Here we meet again the ambiguity (mentioned in §3.1) of what we might mean by 'Bohr's theory' or 'Bohr's research programme'. Only rarely do we mean the particular theory Bohr had; more commonly we mean the ideas of the community committed to following Bohr (e.g. we don't care about Bohr's fifth postulate, on p.875 of his trilogy). But if we interpret Smith as talking about what Bohr himself was committed to, and if Bohr himself was strongly committed to the 'CP' (as Smith interprets it), then Smith's statement isn't wrong. It's just misleading, because (a) it uses the term 'correspondence principle' in a different way to much of the literature (both past and present literature), and (b) it suggests widespread community commitment, since we assume that Smith cares about more than what Bohr alone committed to.

Suppose we accept that, for Bohr at least (and perhaps Kramers), there was genuine commitment to the 'CP' in the form where it can be used to calculate spectral line intensities.[17] Leaving aside the question of why we would care about this inconsistency given how few scientists committed to the relevant propositions, and how unsuccessful it was, do we even *have* inconsistency here? Of course, we have use of propositions taken from CED along with Bohr's postulates which are inconsistent with CED. Now, Bohr stated explicitly that CED ultimately needed to be changed ('generalized') as I already discussed in §3.2. Thus even if he did make use of CED as a whole along with his new postulates about atomic behaviour that wouldn't amount to an inconsistent *belief* set. Instead, Bohr believed that CED was strictly speaking false, but still worthwhile using due to its approximate truth, and its potential to remain true or approximately true in various contexts of application (which might include certain aspects of atomic behaviour). But it remains possible that Bohr was nevertheless *using* inconsistent

---

[17] Alternatively we could talk in terms of a 'correspondence principle' and then further assumptions committed to by Bohr and Kramers about why we should expect the connection between frequencies and intensities at large n to continue when we move to small n. Whether we talk about 'a different version of the CP' or 'the same CP with further assumptions added' is purely linguistic: the lessons vis-à-vis inconsistency remain the same.

propositions together, and this might raise the question of how he avoided explosion via ECQ.

To repeat the point made in §3.2.3, reasoning with inconsistent propositions is not significantly different from reasoning with propositions at least one of which one knows to be false (e.g. reasoning with approximations, idealizations, etc). One cannot have full confidence in any inference one makes in such circumstances, so one is forced to judge one's inferences one at a time, empirically and/or conceptually. Logical explosion via ECQ can only take hold if one first derives explicit contradictories—but then one will obviously notice this and stop making inferences! Scientists will (of course) recognize that any proposition inferred from a contradiction or contradictories will not be (and should not be) taken seriously by the scientific community, even if the inferences used are all unquestionably truth-preserving. The genuine possibility of inferring anything one wants (or doesn't want) via ECQ just does not arise without engaging in the most obviously illegitimate reasoning, from a scientific point of view. In such circumstances there is nothing wrong with the inference rules themselves, of course, but the underlying point of using those inference rules is lost.

But in fact, in the end, it isn't even clear that Bohr and Kramers were reasoning with inconsistent propositions here. As Smith himself puts it, Bohr used 'results that were otherwise justified only by mutually inconsistent accounts of the source of spectral emission' (Smith 1988: 442). That is, Bohr used assumptions about the frequency of spectral lines based on his postulates, along with assumptions about the relationship between frequency and intensity that were *justified* only by various elements of CED. But then, by Smith's own account Bohr was not *using* inconsistent propositions, but rather using a consistent set of propositions which were originally derived from *other* propositions which are inconsistent. The original *justifications* are inconsistent (Bohr's postulates and CED), but not the propositions Bohr ultimately makes use of to predict the intensities of spectral lines. The inconsistency is in the 'mutually inconsistent representations from which those hypotheses [the ones Bohr reasons with] are derived' (Smith 1988: 441), namely, CED and Bohr's postulates.[18]

---

[18] Is Smith supported by Lakatos, as he claims? Lakatos (1970a: 114) does say that the principle *led* to results concerning the intensity of spectral lines. But it's not clear that Lakatos has Smith's understanding of the 'correspondence principle' in mind, as something distinct from the correspondence 'theorem' (as Smith has it), or that the correspondence principle was in some sense the *cause* of the inconsistency for

3.3.2 The adiabatic principle

Two systems are adiabatically related if the second can be achieved by taking the first and changing a certain parameter infinitely slowly and smoothly (see Scerri 1993). For present purposes there is no need to go into the details of the principle; it is enough to note that it was a way to proceed from knowledge of a familiar quantum system to knowledge of a new quantum system. Using this principle, Pauli derived a contradiction in 1926.

To appreciate this particular inconsistency a little background is required. First of all it should be noted that postulate P2 (or, if you prefer, P2*), sometimes called the 'quantum condition', was radically updated by Sommerfeld. In 1916 he achieved a much celebrated explanation of the fine structure of the hydrogen spectrum (how each line splits into further lines if one looks through a high resolution spectroscope) by introducing a second quantum number which restricted the possible eccentricities of the (elliptical) orbits, and by introducing relativistic effects. His equation can be expressed as,

$$E_{n,k} = \frac{hR}{n^2}\left\{1 + \frac{\alpha^2}{n^2}\left(\frac{n}{k} - \frac{3}{4}\right)\right\} + \cdots$$

where k is the 'second quantum number' in addition to Bohr's principal quantum number n, and $\alpha \approx 1/137$ (as before, h is Planck's constant and R the Rydberg constant). The dots stand for terms in α^4, α^6, etc., so they can be neglected.

Later that year Sommerfeld went still further, introducing third and fourth quantum numbers s and m. These quantized the possible orientations of the stationary states in the presence of an external electric or magnetic field (that is, the angle which the plane of the orbit makes with the incoming field). Some significant success was achieved in explaining both the Stark effect—the splitting of the spectral lines due to the presence of an external *electric* field—and what was called the 'normal' Zeeman effect—the splitting

---

Lakatos. In fact Lakatos refers to the correspondence principle as an 'ad hoc stratagem', which helps to 'conceal the graft [mixing of principles]', and in so doing 'reduce[s] the degree of problematicity of the programme' (1970a; 144). The theory was already inconsistent in Lakatos's eyes, and the introduction of the correspondence principle helped to accommodate the inconsistency somehow. Lakatos probably identified the inconsistency as the one discussed in §3.2.3, above, as Bartelborth (1989) suggests.

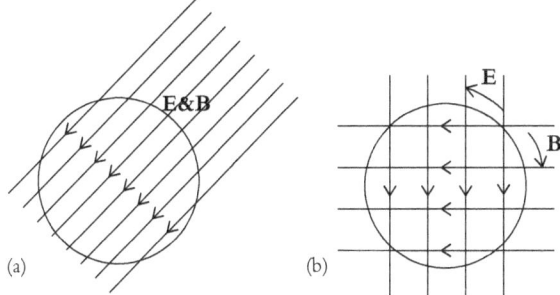

Figure 3.1 (a) Electric and magnetic fields coming at a hydrogen atom from the same direction; (b) the directions of the electric and magnetic fields are turned infinitely slowly and smoothly (as mathematically conceived) in opposite directions until they cross each other at right angles

of the spectral lines due to the presence of an external *magnetic* field. Thus s was called the 'electric quantum number' and m was called the 'magnetic quantum number' (see Pais 1991: 199).

Restrictions were put on the possible values of these extra quantum numbers. The second quantum number k was made to be less than or equal to n, for example. Of the other restrictions, the important one for present purposes is that m could not be zero. If m were allowed to be zero this would refer to an 'orbit' called a 'pendulum orbit'. In this case the electron would head *straight towards* the nucleus, and then (inexplicably, since there would be no slingshot effect) head away from the nucleus again on the same line.

Now Bohr and Sommerfeld were realists about the orbits, so for them such pendulum orbits were impermissible.[19] To make them impossible within the theory it had to be stipulated that the magnetic quantum number m could never be zero: '$m \neq 0$'. But in 1926 Pauli showed that this was incompatible with the adiabatic principle.[20]

First Pauli considered the hydrogen atom with electric and magnetic external fields coming from the same direction as in Figure 3.1(a). This gave certain allowed orbits, decided by giving the appropriate values for the

---

[19] Their realism explains their developments of the theory: consideration of elliptical orbits, relativistic adjustments to orbital trajectories, etc. As Darrigol puts it, 'Bohr and Sommerfeld excluded the value m=0 on the grounds that the corresponding orbit is adiabatically connected to an orbit passing through the nucleus' (1992: 188). See also Lindsay (1927: 413).

[20] In what follows I draw on Mehra and Rechenberg (1982: 507–9).

quantum numbers n, k, m, and s in the quantum condition. Then Pauli considered an adiabatic change to the system, rotating the electric and magnetic fields in opposite directions as in Figure 3.1(b). Following the rules of the adiabatic principle, it turned out that by doing this a system could be achieved theoretically for which the magnetic quantum number m was equal to zero. But, as noted above, $m \neq 0$ was stipulated as a necessary feature of the quantum condition. As Pauli put it, 'the "allowed" orbits, s = 0, m $\neq$ 0, pass over into the "forbidden" orbits, m = 0, s $\neq$ 0.' He goes on: 'An escape from this difficulty can be achieved only by a radical change in the foundation of the theory' (Pauli 1926: 163–4). The necessary change was so radical, it would seem, since both $m \neq 0$ and the adiabatic principle had been integral parts of OQT since at least 1918. In other words, it appeared there had been an inconsistency lurking for a very long time in some of the most fundamental assumptions committed to by the community.

What we learn from this inconsistency, and how much it matters, will depend (for one thing) on the way in which the community 'committed' to the primary propositions in play (Sommerfeld's quantum condition and the adiabatic principle). For example, I've said that $m \neq 0$ was forced upon Bohr and Sommerfeld because they were realists about the orbits, but what did others make of this suggestion? Simply put, the fact that Sommerfeld's introduction of elliptical and relativistic orbits was so explanatorily and predictively successful convinced nearly everyone, since these developments were based on a realistic picture of events.[21] Some did object to taking certain *parts* of the model realistically: for example, Debye (who independently explained the Zeeman effect in the same way as Sommerfeld in 1916) was sceptical about a realistic 'spatial orientation' interpretation of the electric and magnetic quantum numbers s and m. When Stern and Gerlach tested the explanation of the Zeeman effect in 1921 Debye wrote, 'But you surely don't believe that the [spatial] orientation of atoms is something physically real; that is [only] a prescription for the calculation, a timetable for the electrons' (quoted in Friedrich and Herschbach 2003: 56). Presumably then, for Debye, allowing for '$m = 0$' would not be a problem, as it wouldn't mean that electrons were *really* going through the nucleus. However, the results of Stern and Gerlach's experiment went firmly against

---

[21] See Vickers (2012) for evidence of this in the form of quotations from members of the scientific community.

Debye's scepticism (Friedrich and Herschbach 2003: 56). If not the whole community, then at least a large proportion would not have allowed for an 'orbit' which went straight through the nucleus, on realist grounds. Thus at least a large proportion of the community would have made a serious commitment to Sommerfeld's '$m \neq 0$' during those years.

In addition evidence can be mounted that there was a serious commitment to the adiabatic principle. Paul Ehrenfest brought it to the attention of the scientific community in a 1917 paper. At first it wasn't clear how it could be used in specific calculations, but this soon changed. As Brown notes, 'With its help, he [Ehrenfest] was able to determine quantization rules for a wide range of systems, given only a rule for one of them' (1992: 402). Sommerfeld, at first sceptical, wrote of the predictive power of the principle in 1919 (see Klein 1985: 291). And its explanatory value was widely accepted: Jammer writes, 'The adiabatic principle ... revealed the mystery of the quantum conditions' (1966: 101). The 'wide and ingenious use Bohr had made of the adiabatic principle' might also be noted (Bergia and Navarro 2000: 28). In short we have prediction, explanation, ubiquitous use, and widespread endorsement from the physics community. It is surely right to say that, 'from 1917, Ehrenfest's adiabatic principle remained strictly linked with the development of OQT' (Bergia and Navarro 2000: 10). In addition it should be noted that the community explicitly accepted that the adiabatic principle applied to the subject of Pauli's calculation, the hydrogen atom. The conditions which had to be satisfied for the adiabatic principle to be applicable to a system were very carefully specified, as follows:

(i) The introduction of external forces should not alter the degree of periodicity of the system;
(ii) The system should be simple-periodic or multiple-periodic (not aperiodic).

The hydrogen atom satisfies these conditions (as explained in Scerri 1993: 52).

So the community certainly committed to the relevant propositions, but what exactly was the character of the commitment? Was there a doxastic commitment, or was it more like the case of Bohr's postulate (P2), something *used* for convenience but never considered even a candidate for the explanatory *truth*? Of course, with (P2) Bohr always knew that he'd idealized, that in the final analysis one would want to make the orbits elliptical and consider relativistic effects. But with Sommerfeld's development of the quantum condition the relevant de-idealizations had already been applied. A final possibility was to somehow incorporate electron spin, in the classical sense. But it seems highly unlikely that any account of electron spin would adjust Sommerfeld's condition in such a way that the stipulation $m \neq 0$ would become unnecessary. No scientist could reasonably dismiss the Pauli contradiction as an artefact of the fact that electron spin had not yet been painted into the picture.

Was the adiabatic principle meant to be approximate in some way? Was something intentionally left out of the principle for pragmatic purposes in a way analogous to Bohr's postulate (P2)? The only possibility seems to be the stipulation that the parameter of interest be altered 'infinitely slowly and smoothly'. One might object that, since 'infinite' smoothness is an impossibility—a clear idealization—one can ignore the contradiction on the grounds that the final state Pauli derives, with $m = 0$, could never arise in practice. But this would be to misunderstand the adiabatic principle. It tells us which other quantum systems are possible (given an initial quantum system). It doesn't also promise us that it is possible to carry out the transformations experimentally.

Thus we get some sense of why, upon deriving the contradiction, Pauli exclaimed that a 'radical change in the foundation of the theory' was necessary. In this case we have widespread community commitment not just to *using* the relevant propositions, but commitment to the propositions as the best bet (at that time) for the explanatory truth of various atomic phenomena. In the terminology of Chapter 2, the propositions in question are clearly historically relevant—widespread community commitment and endorsement—and pointedly grouped—used together to explain a unified range of phenomena (especially concerning the hydrogen atom). In this situation, Pauli quite rightly noted that the contradiction could not be ignored on any of the obvious grounds, as an artefact of an idealization assumption or a speculative ampliative inference in the derivation (say).

What, then, do we learn from this inconsistency? For one thing it is clear that an inconsistency can lay buried in the complications of various propositions for many years, especially when generalizations are involved (such as the adiabatic principle). There is no reason why Pauli could not have made his derivation as early as 1917 or 1918: all of the necessary propositions were known to him, and others, at that time. It was simply a matter of applying them to the hydrogen atom in the right way. But it is clear why it did in fact take so long: there are literally hundreds of ways in which one might apply electric and magnetic fields to a hydrogen atom and then consider a possible adiabatic transformation of those fields.[22] It may be that the particular way Pauli did it was the only way in which to derive a contradiction. And in addition, unless one is dedicating one's research to hunting down inconsistencies, it isn't clear why the type of investigation Pauli made is an especially interesting or important investigation. This is something we will come across again in future chapters: the question which leads to a contradiction can sometimes be a question which on the surface isn't particularly interesting, a question which you would be unlikely to acquire funding to investigate.

This case also stands as a concrete example of how ECQ need not pose a *special* problem when we reason with inconsistent propositions. The only way to derive *anything we want* is to first reach explicit contradictories, and then manipulate those contradictories (as already noted). It should be obvious from the details of this case that that just isn't going to happen in real science, and certainly not by accident! The case is analogous to the Goldbach case I discussed in §3.2, except now we have a real-life scientific example. There are numerous ways to reason with the propositions, and it may be that *very* few of these actually result in deriving contradictories. And when contradictories *are* reached, one doesn't *continue* and derive arbitrary propositions: one stops what one is doing and declares that something in the original assumption set needs to be changed (as Pauli did). Feyerabend's words from Chapter 1 resonate here: science really does handle contradictions in a less simple-minded fashion. The contradictories tell us that at least one of the propositions in our original assumption set is false, but from here the lessons for the trustworthiness of inferences are no different from those for reasoning with idealizations, approximations, simplifications, abstractions, and so on.[23]

---

[22] Cf. the hundreds of ways in which one can perform universal instantiation from Goldbach's conjecture.

[23] Harman (1986) suggests that one should continue to use classical logic when reasoning with inconsistent assumptions, and just try to avoid absurd results in a commonsense way. This is something like what I am endorsing here, but Harman doesn't quite see just how justified his approach is when it comes to real-life scientific examples. He writes 'The danger is that, since inconsistent beliefs logically imply anything, if one is not careful, one will be able to use this fact to infer anything whatsoever' (1986: 16). But as the case currently under discussion shows clearly, this isn't a 'danger', and one does not need

The lessons for doxastic commitment may be different. Perhaps nothing needs to be changed right away for the purposes of making inferences, on the grounds that logical explosion doesn't really threaten as already explained. But when we ask ourselves what we want to believe, something really does need to be changed, and it may be hard to see what that could be. If all of the propositions in question have enjoyed some measure of explanatory or predictive success, say, then in rejecting any one of them we also wave goodbye to the corresponding success. So the question might be, which of the propositions in the Pauli inconsistency can be ejected with minimal damage?

Now this question didn't really arise in a serious way in the relevant scientific history, because by 1926 Heisenberg's matrix mechanics and Schrödinger's wave mechanics were changing the scientific landscape dramatically. But it is interesting to note that, absent these new developments, the consistency of the theory could have been recovered without the 'radical change' that Pauli supposed was necessary. In fact, a suggestion made by R. B. Lindsay in 1927—although not motivated by Pauli's result—would have done the trick. First, $m \neq 0$ should be ejected from the theory. This is an obvious candidate for ejection in a sense, because it is not in itself doing any explanatory or predictive work, and it is an independent part of Sommerfeld's quantum condition. Of course it was doing *conceptual* work, because removing it seems to allow for impossible 'pendulum orbits' right through the nucleus. One option would be to do away altogether with the realist attitude to orbital trajectories of electrons, in the style of Debye (above). But of course, this would then suggest that the whole picture provided by Bohr's postulates has nothing to do with reality, but is just an instrument for calculation. Even Debye never wanted this. But a less radical option was provided by Lindsay (1927):

> [T]he idea of the passage of the electron through the nucleus may be distasteful to some. There is a possible way of avoiding this, namely, by the introduction of a repulsive force (in addition to the inverse square attractive force) operative only in the immediate vicinity of the nucleus. (Lindsay 1927: 415)

---

to be 'careful' to avoid explosion. The concern Harman articulates here is ubiquitous in the literature, but unfounded.

Thus the electron would not travel through or collide with the nucleus, but would bounce back with the newly introduced repulsive force. We would end up with something very like the strong force in modern nuclear physics.

3.4 Conclusion

I made a major and no doubt controversial claim in Chapter 2: that science—at least *inconsistent* science—is not nearly as uniform as philosophical literature usually assumes. One consequence, I suggest, is that philosophy of science which talks in terms of 'theories' in the history of science about which we can say something quite general is treading on very dangerous ground. The more specific claim in the present context is that philosophy of science which tries to say something quite general about *inconsistencies* in the history of science is in danger of doing an injustice to the real history of science. All inconsistencies are unequal, I claimed. Well of course they are, but the claim is that the differences are more substantial than is assumed by literature making very general claims concerning 'how to reason from inconsistent theories', for example. And this to the extent that such literature can be justifiably described as based on a false premise.

I've only looked at one 'theory' so far—Bohr's theory of the atom—but already there is some evidence to support my claims. Five inconsistencies have been introduced, and there are important differences between them concerning what I have called 'historical relevance' and 'point of grouping'. This already suggests that any general theory of how to reason from inconsistencies in science is not going to be helpful. For one thing, it isn't clear that we need any account of how to reason from inconsistencies in the first four cases, because it's not clear in each case that anybody *was* reasoning from inconsistent propositions. The extent to which scientists were reasoning from inconsistent propositions is a complex story in each case. For example, in the first case it depends on the sense in which a principle prohibiting certain forms of discreteness in science was 'present' in 1910s physics, and being acted upon explicitly or implicitly by members of the scientific community. In the second case it depends on the sense in which scientists interpreted the 'quantum transitions', and the sense in which the community was committed to worldline continuity of particles. The third case is rather different: in this case it depends (for one thing) on the grounds on which we might draw a distinction between CED and certain consequences *of* CED which were sometimes used alongside Bohr's postulates. In the fourth case it depends on the sense in which the correspondence principle was playing a role in Old Quantum Theory (OQT), and the sense in which it would be appropriate to introduce the principle in a form which permits us to calculate spectral line intensities for small quantum number n. In this case an extra complication is that different members of the relevant scientific community clearly took very different attitudes to the correspondence principle.

It is revealing that the final inconsistency is the only one where a contradiction was actually derived, and presented to the rest of the community as a serious problem demanding resolution. Of course, those in the community who despaired of Bohr's theory from the start had long seen it as riddled with serious problems.[24] But with Pauli we find the first real example of a scientist committed to working with the relevant propositions in a serious way calling for 'a radical change in the foundation of the theory'. This demonstrates vividly the different and more serious character of the Pauli inconsistency in comparison with the other four inconsistencies.

Given that this fifth inconsistency was taken so much more seriously than the others by the relevant scientific community, it is remarkable that *it* isn't the focus of internal inconsistency for 'Bohr's theory'. It is worrying that we have such an entrenched 'received view' in the literature that 'Bohr's theory of the atom' is internally inconsistent, but without any appreciation of the myriad differences between the different inconsistencies presented here, and without any mention of the Pauli inconsistency. An underlying assumption in much of the literature appears to be that Bohr's theory has unambiguous content as a set of propositions, and that 'the theory' is inconsistent in the straightforward sense that this set of propositions is formally inconsistent. The analysis I have provided in this chapter suggests that this view does a major injustice to the complexities of science as it is really practised. For

---

[24] E.g. Alfred Landé recalls scientists in 1914 stating, 'If it's not nonsense, at least it doesn't make sense.' And Ehrenfest stated in 1913, 'If this is the way to reach the goal I must give up doing physics.' For these and further quotations, see Pais (1991: 152–5), Klein (1985: 278), and Jammer (1966: 86f.). Of course, one doesn't need to appeal to inconsistency to explain such sentiments: one can appeal to other 'conceptual problems', such as incompleteness, ad hoc-ness, and vagueness (see Darden 1991: 201ff. and Newton-Smith 1981: 227ff.).

example, one might wonder whether it even makes sense to ask whether the correspondence principle is a 'part of the theory', given that the individuals in the relevant community had many different interpretations of the principle, and different interpretations were used in different contexts, with a variety of measures of success.

It is these sorts of difficulties that the 'theory eliminativism' method is designed to handle. For example, the disagreement between Shapere (1977) and Smith (1988) over whether the correspondence principle was a 'part of the theory' (in a form where it could tackle spectral line intensities) is dissolved.[25] All that matters is what we learn from the inconsistency Smith (1988) considers, what lessons there are for how science works. I have argued that the inconsistency isn't especially interesting, but this isn't because the principle isn't a 'part of the theory'. It is because of who really committed to the principle (in Smith's sense), and how they committed, and the particular way in which CED was used to calculate spectral line intensities. Theory eliminativism forces us to ask these deeper questions about why an inconsistency does or doesn't matter to philosophy of science. Similar benefits come in the other cases: for example, in the first case we don't ask whether the continuity principle in question is/was a 'part of the theory', but ask what we learn about how science works from the fact that this principle is contradicted by Bohr's postulates.

Many of these conclusions are negative, in the sense that they are about how philosophy of science shouldn't be done. But of course, they are also about how it can be done better. In addition, there are positive lessons about how science works in the face of certain inconsistencies. An important distinction between doxastic commitment and a commitment just to *use* inconsistent propositions was introduced. In the latter case, I argued that ECQ does not present anything like the problem that is usually assumed in the literature: scientists are justified in reasoning with inconsistent propositions for precisely the same reasons that they are justified in reasoning with idealization and approximation assumptions. A set of inconsistent assumptions can be approximately true in the strongest possible sense: when every

---

[25] Cf. Shapere's concerns: '[T]he notion of theory is today in a worse state than ever.... There is today no completely—one is almost tempted to say remotely—satisfactory analysis of the notion of a scientific theory.... [There is a] lack of precision, in usual discussions, as to what is to count as "(part of) a theory"' (Shapere 1969: 124–5 and 139). If he had employed theory eliminativism he could have completely forgotten about these 'problems'.

assumption is true except for one, which is itself approximately true (cf. the Goldbach inconsistency if only *one* even number in the set of all integers is not the sum of two primes).

The case of doxastic commitment can only arise when scientists don't know about the inconsistency. It is the fact that the community didn't know about it that made the Pauli case so surprising, and so serious. How should science proceed in such a case? I suggested a strategy in the previous section: look back over the inconsistent propositions, and consider what can be dropped, or replaced, with minimal damage to accumulated scientific success. In an ideal world there will be at least one assumption that isn't really doing any scientific work at all, that can be omitted without loss. This is a general claim about how to reason in such cases, but it is only a guideline. In each case it will depend on how exactly propositions are 'doing scientific work', how they are unified with the other propositions, and the ways in which they might be replaced. Some further lessons about how this might be done will emerge in the case studies to follow.

Finally here, let me highlight a distinction between two possible attitudes to the question 'So *was* Bohr's theory inconsistent or not?' I said in Chapter 2 that I would leave this question aside. One can eliminate *theory* and just ignore this question, taking no stance on the question of whether there is a fact of the matter whether 'Bohr's theory' was inconsistent. One can just argue that the important questions about inconsistency in science can be asked and answered without taking a stance on the question of whether there is a fact of the matter. What I said in Chapter 2 is compatible with this stance: I made no claim that 'theories don't exist', or even that there isn't some specific set of theoretical constituents which constitute a given theory.

However, the analysis presented in this chapter goes at least some way towards showing that the question 'Is the theory inconsistent?' really is a bad question, based on a false premise. When we look to the history we find various different propositions used in different contexts, and committed to, and used, in a variety of formulations, in various different ways, by different members and sub-communities of the relevant scientific community. Various issues concerning historical relevance and the 'point of grouping' of different sets of propositions arise. In these circumstances, why would we think that there is something the theory 'was' or 'is'? The complexities of real scientific history resist any answer to the question 'what is the theory?'

This is hardly to give up on the question because it is 'too hard'. It is to realize that the way human beings conceptualize things and use language just does allow for questions that at first seem sensible, but are in fact bad, conceptually confused questions, based on false premises.

As I said, there is no need for me to make this claim to motivate theory eliminativism as a method. But perhaps it is worth flagging up at this stage the possibility that the motivation for theory eliminativism is not just pragmatic, but also in a sense ontological. Evidence for this ontological motivation will accumulate gradually as we go through the case studies: the historical and scientific messiness of Bohr's theory is representative of science generally, I submit. Even a theory as apparently well-defined as CED suffers from this problem. And CED, just as Bohr's theory, has been the target of an inconsistency claim, as we will see next.

4

Classical Electrodynamics

4.1 Introduction

Classical electrodynamics is the most recent example of an 'internally inconsistent theory' to be added to the mix. Mathias Frisch originally made the claim in a 2004 paper (Frisch 2004), but then reasserted it and defended it in much greater detail in the first three chapters of his book *Inconsistency, Asymmetry, and Non-locality* (Frisch 2005a). The claim was fiercely opposed by Fred Muller (2007) and Gordon Belot (2007), but here I show that things are not so clear. This case serves as a prime example of how, even in our most recent literature, thinking in terms of 'the theory' can cause miscommunication and hinder productive debate.

In §4.2, I outline some important features of the theory and introduce some conceptual complications which accompany one of these features, namely, the Lorentz force equation. In §4.3, we are taken to Frisch's inconsistency claim and how he manages to derive a contradiction. In §4.4, Frisch's claims are defended against the recent criticisms of Muller and Belot, before I assess the significance of the Frisch inconsistency in §4.5. In §4.6, I conclude the chapter.

4.2 Features of the Theory

The theory in question is Maxwell–Lorentz classical electrodynamics (CED), the same theory discussed in the previous chapter as inconsistent *with* Bohr's theory of the atom. It explains electromagnetic (EM) phenomena by describing interactions between microscopic charged particles and EM fields. The ontological distinction between the particles and the fields divides the laws of the theory in two. On the one hand we have the Maxwell equations (MEs) which tell us how particles give rise to and affect fields. On the other hand we have the Lorentz force equation (LFE) which tells us how fields affect particles. The interaction of these laws is in part governed by the postulate of energy conservation: it is assumed that for a given system of particles and fields the sum of the kinetic energies E_k of the particles and the field energy E_f remains constant over time. Since energy is transferred between particles and fields, this means that for any increase in particle energies there is a corresponding decrease in field energy, and vice versa. For a closed system we can write,

$$\Delta E_k + \Delta E_f = 0.$$

If energy is leaving or entering by crossing an imaginary surface enclosing the system the conservation equation becomes,

$$\Delta E_k + \Delta E_f = E_{\text{over surface}}.$$

Strictly speaking this is meant to stand for any given system, and for any period of time however small.

The LFE will be central to this chapter, so I will say a little more about it here. As I said, it tells us how fields affect particles. It is often presented as

$$\mathbf{F} = q(\mathbf{E} + \mathbf{v} \times \mathbf{B}),$$

with the vector quantities $\mathbf{E}$ and $\mathbf{B}$ denoting the electric and magnetic field properties at the point (or points) of the particle, and q, $\mathbf{v}$, and $\mathbf{F}$ denoting, respectively, the charge on, velocity of, and (Lorentz) force experienced by the particle.[1] The idea is that we already know $\mathbf{E}$, $\mathbf{B}$, q, and $\mathbf{v}$, and we use the equation to determine $\mathbf{F}$. However, there are some important complications hidden in the interpretation of $\mathbf{E}$ and $\mathbf{B}$, and the terminology in the literature can be misleading.

Consider a single charged particle in uniform motion at time t_o. Which sources of field must the LFE take into account? The first and most obvious are the *external* fields, those which are reaching the particle at time t_o from other charged particles. It is fundamental to CED that external fields affect the particle, so they should play a part in the LFE. To make things clear,

---

[1] Following Frisch and Belot, I take it that I may ignore the relativistic generalization of CED for the majority of the present discussion. However, see §4.4.2, this volume.

these fields are sometimes distinguished from other fields by the subscript 'ext'. We can write,

$$\mathbf{F}_{ext} = q(\mathbf{E}_{ext} + \mathbf{v} \times \mathbf{B}_{ext}),$$

which tells us the force due purely to external fields. The total force is the combination of the external force and the so-called 'self-force': $\mathbf{F}_{tot} = \mathbf{F}_{ext} + \mathbf{F}_{self}$. It remains to determine what '$\mathbf{F}_{self}$' stands for. If external fields cause the external force, what are the *self*-fields which cause the *self*-force?

The most obvious self-field is the familiar Coulomb field. In its most common, introductory form it is a purely electric field which surrounds any charged particle, diffusing out into space isotropically and pointing (in its vectorial representation) directly away from the particle. It is proportional to the charge of the particle and inversely proportional to the square of the distance from it, so we can write,

$$\mathbf{E}_{coulomb} = k\frac{q\hat{\mathbf{x}}}{r^2}$$

(where $\hat{\mathbf{x}} = \mathbf{x}/|\mathbf{x}|$ and k is a constant). This field can be thought of as attached to the particle, as it accompanies it everywhere it goes. No energy is carried away from the particle; arrows pointing away from the particle which represent the field at a point don't represent the actual movement of anything, but just give the direction another, similarly charged particle *would* be pushed *if* one were placed there.[2] It acts on *other* particles as an *external* field, but here we are asking whether it should play any part in the LFE, where we consider the effect on the particle which plays host to that field.

In a classical frame of mind we might now suppose that the particle must either be a point particle or be extended. (There are questions about whether either conception is ultimately coherent, but for now let's suppose that they are, and continue asking the question how self-fields and self-force might be added to the picture.[3]) If it is a point particle, then of course at the point of the particle $r = 0$, so the Coulomb field there is going to be

---

[2] The flux over the surface of a sphere enclosing the particle is non-zero, and 'flux' comes from the Latin for 'flow', so one could be forgiven for imagining something flowing away from the particle. However, 'flux' should only be taken to mean 'flow' when the vector field in question is representing movement. Compare the contours of a hill to the blowing of the wind.

[3] See Frisch (2005a), especially ch. 3, for further details. See also §7.4, this volume.

infinite in magnitude. Even for the extended case, the field at the surface of the particle is going to be very large, since r will be very small. However we might find some comfort in the fact that, for a stationary particle (or indeed one in an inertial frame) the Coulomb field will surround the particle symmetrically. As such any forces will all cancel each other, to give zero net force. So in this case at least, the Coulomb field should *not* play any part in the LFE.

However, when the particle is accelerating this symmetry breaks down. One might imagine the Coulomb field lagging behind its particle and constantly trying to catch it up. In this case it looks like the particle *will* experience a force due to its own field. But how can this be written into the LFE? The details will depend essentially on the structure of the particle, whether it is a point, a rigid sphere, a rigid shell, a dumbbell, a non-rigid body, or whatever. If the particle is not a point, then the difference between the field strength on one side of the body and the other is going to play a part, adding to the complications. So although a particle's own Coulomb field is relevant to the Lorentz force on a particle in certain circumstances, it isn't obvious how to include it in the LFE.

The term 'Coulomb field' is usually reserved for the isotropic, electric field which surrounds a charged particle when observed from its own rest frame. But when we consider the particle to be accelerating, and we introduce relativity, the field is in general neither purely electric nor symmetric. Authors (Duffin 1990; Jackson 1999) often use the term 'static field' instead for the field which is in some sense 'attached' to the particle (so the Coulomb field is a special case of a static field). The static field provides the contrast for another type of self-field, the 'radiation field'.

As already discussed in the previous chapter, when charged particles accelerate—at least *outside* of atoms—they emit radiation. In the classical theory this radiation is characterized as an EM wave which falls off in intensity with $1/r$ from its host particle. It dominates the static field at large distances from the particle (since the latter goes with $1/r^2$), and is taken to exist independently of the static field. Thus it cannot be dismissed as a disturbance in the static field already discussed; instead it takes on a life of its own, independent of *any* medium. The things that 'wave' are electric and magnetic field properties which are directed perpendicular to both (i) the direction of travel, and (ii) each other. Since this radiation is caused by the particle, and is not a wave in a medium but carries its own fields with it, so to speak, it can fairly be described as an emitted *field*. And it often is so described. Muller (2007: 259) writes of

The energy radiated by the moving charge, via its emitted electro-magnetic field, often called the *self-field*, $\mathbf{E}_{self}$ and $\mathbf{B}_{self}$.

Muller here apparently uses the term 'self-field' to refer to the radiation field alone, but it is much more usual to use this term to refer to *all* fields which are not external, radiation *and* static fields. Feynman et al. (1964: 28-5) and Belot (2007: 271) are cases in point, and I will follow their lead.[4] What is significant is that in all cases the emitted radiation is referred to as self-*field*. As Muller says, energy leaves the particle in the form of this radiation. Energy is conserved, the kinetic energy of the particle is decreased, and the particle's acceleration is reduced—so the particle does experience a self-force.

We have here another type of self-field affecting our particle, but again it is far from obvious how to write this into the LFE. To start with, the particle is taken to experience a force because it loses energy, not because it interacts with the radiation field's $\mathbf{E}$ and $\mathbf{B}$ components. If we do wish to include the radiated field in the LFE *as field*, then since it falls off with $1/r$ we experience essentially the same difficulties we saw with the $1/r^2$ Coulomb field when r becomes very small. How the field affects the particle will depend on the particle's structure.

However, it is in fact a fallacy to speak of the radiation field causing the host particle to experience a force. This attitude is prevalent in the literature; for example Griffiths (1999: 465) writes, 'The radiation evidently exerts a force ($\mathbf{F}_{rad}$) back on the charge—a *recoil* force, rather like that of a bullet on a gun'. But the recoil on a gun really has the same, common cause as the ejection of the bullet, namely the rapidly expanding gas inside the chamber. Equally, the 'recoil' of a particle really has the same common cause as the emission of radiation. Feynman writes,

With acceleration, if we look at the forces between the various parts of the electron, action and reaction are not exactly equal, and the electron exerts a force *on itself*. (Feynman 1964: 28-5, original emphasis)

---

[4] In fact the distinction between 'static' and 'radiation' fields cannot be made in the relativistic generalization of the theory, where one works with the field tensor F rather than the vector fields E and B.

In other words, the self-force is caused by the *static* fields of different parts of the electron affecting other parts as *external* fields (assuming an extended particle), and when the particle accelerates these forces don't balance. And if the particle loses energy in this way, that energy has to *go* somewhere. The radiation field is the manifestation of this loss of energy.[5]

Where does all this leave us? Our characterization of the LFE so far is $\mathbf{F}_{ext} = q(\mathbf{E}_{ext} + \mathbf{v} \times \mathbf{B}_{ext})$, plus an extra self-force effect when the particle is accelerating. We have two options for characterizing this self-force:

(1) look to the energy emitted as radiation, and calculate the force on the particle using energy conservation;
(2) introduce a model of the particle, and calculate the force on different parts of the particle due to the static fields of other parts.

In fact, neither option leads to a satisfactory result. Griffiths and Jackson explore both possibilities. The first option (1) leads to the Abraham–Lorentz equation, which can be written in its non-relativistic form as,

$$\mathbf{F}_{rad} = \frac{2q^2}{3c^3} \dot{a}$$

(Griffiths 1999: 467; Jackson 1999: 748). Unfortunately this equation doesn't follow straightforwardly from the core of the theory. Jackson notes that the derivation is 'certainly not rigorous or fundamental' (1999: 750). And in any case it carries with it severe conceptual difficulties. If it *did* follow rigorously from the theory, the conceptual difficulties attached to it might be used to make a quite different inconsistency claim to the one in question in this chapter.

The other option (2) makes $\mathbf{F}_{self}$ a function of the particle's self-fields, $\mathbf{E}_{self}$ and $\mathbf{B}_{self}$, rather than its acceleration. It is usual, as noted above, to introduce a model of the particle in question and calculate how it interacts with its own fields. Now if we think of different parts of the particle

---

[5] A little later in his book Griffiths tells a similar story (1999: 472), and so too, briefly, does Belot (2007: 269).

affecting each other through *external* fields, then the equation for $\mathbf{F}_{self}$ will mirror the equation for $\mathbf{F}_{ext}$:

$$\mathbf{F}_{self} = q(\mathbf{E}_{self} + \mathbf{v} \times \mathbf{B}_{self}).$$

$\mathbf{F}_{self}$ is just the *external* force on a given *part* of the particle due to other parts of the particle. When we add to this the *other* external forces we get a 'total-LFE': $\mathbf{F}_{tot} = q(\mathbf{E}_{tot} + \mathbf{v} \times \mathbf{B}_{tot})$. However, what we wanted to know was how fields affect particles *as a whole*. To know the *net* force a particle exerts on itself we need to know the structure of the particle, because only then can we add up all the $\mathbf{F}_{self}$ terms for all the different parts. This is awkward: it is notoriously difficult to reconcile any model of an elementary charged particle with physical 'common-sense', empirical results, and the constraints of special relativity.[6] Even if we ignore these difficulties, any proposed model of an electron is going to be highly speculative. In addition we might wonder whether, in addition to the particle as a whole having an effect on itself, each *part* of the particle has an effect on itself (cf. Griffiths 1999: 472).

It is also possible to work towards such an equation *without* introducing a particle model (see §4.4.2, below). But, as we will see, $\mathbf{F}_{self} = q(\mathbf{E}_{self} + \mathbf{v} \times \mathbf{B}_{self})$ only follows from a *relativistic* version of the theory, and is in any case of no *use* without introducing a model of the particle involved.

It is in trying to come to terms with these difficulties with the LFE, and how they are solved in practice, that Frisch's inconsistency claim is born.

4.3 Frisch's Inconsistency Claim

Frisch states that, '[T]he *core assumptions* of the Maxwell-Lorentz approach to microscopic particle-field interactions are inconsistent with each other' (2005a: 34, emphasis added). He also calls these assumptions the 'fundamental principles' of CED (p. 39). The assumptions he has in mind are the following (p. 33):

(1) There are in fact charged, accelerating particles.
(2) The MEs
(3) The LFE
(4) Energy is conserved in particle-field interactions.

---

[6] Again, Frisch (2005a) is an excellent philosophically minded introduction to these issues.

Prima facie these do indeed seem to be fundamental to the theory. If we are to eliminate *theory*, we might suppose that any inconsistency in these propositions is bound to be a highly important and significant inconsistency for various reasons. However, it is one thing to summarize one's assumptions, and quite another to explain what one means by them. In particular— in light of the discussion in the previous section—we may ask just what precisely Frisch means by the 'LFE'.

On p.27 he introduces it as $\mathbf{F}_{Lorentz} = q(\mathbf{E}_{ext} + \mathbf{v} \times \mathbf{B}_{ext})$. Since only external fields are acting in this equation, one might initially assume that $\mathbf{F}_{Lorentz}$ must stand for the force due only to external fields. This might then be explicitly written into the equation as $\mathbf{F}_{ext} = q(\mathbf{E}_{ext} + \mathbf{v} \times \mathbf{B}_{ext})$. This is apparently confirmed on p.30 where Frisch states, 'The effect of *external* electromagnetic fields on charged particles is given by the Lorentz force law' (2005a: 30, emphasis added). However, reading more carefully, it is clear that Frisch does *not* mean for us to interpret the LFE in this way. For example, on p.33 he writes, 'According to the Lorentz force law, the energy change of a charge is due *only* to the effects of external forces' (emphasis added). It follows that he really construes the LFE as $\mathbf{F}_{tot} = q(\mathbf{E}_{ext} + \mathbf{v} \times \mathbf{B}_{ext})$, that is, the *total* EM force experienced by a charged particle is a function of the *external* fields only.

Now as noted in §4.2 we can write $\mathbf{F}_{tot} = \mathbf{F}_{ext} + \mathbf{F}_{self}$. Accordingly we can split Frisch's LFE into two parts, corresponding to the two different sources of fields in play. We get a purely external LFE, $\mathbf{F}_{ext} = q(\mathbf{E}_{ext} + \mathbf{v} \times \mathbf{B}_{ext})$, and a purely self-field LFE, $\mathbf{F}_{self} = \mathbf{0}$. It emerges that Frisch's LFE asserts, fairly explicitly, that a charged particle experiences no force due to self-fields under any circumstances.

But I said in the previous section that, according to CED, a particle *does* experience a force due to its self-field when it is accelerating, whether we imagine it interacting with its own static fields or imagine it 'recoiling' from the emitted radiation field. In fact this follows from the MEs (2) and energy conservation (4): when a particle is accelerating the MEs tell us that energy is radiated, so by energy conservation the particle loses energy, and by work done it experiences a force, $\mathbf{F}_{self} \neq \mathbf{0}$. In other words we make the following inference:

$$\sim(E_{rad} = 0) \vdash \sim(\mathbf{F}_{self} = \mathbf{0}).$$

But we have just seen that Frisch's LFE tells us that in all circumstances (including when the particle is accelerating) $\mathbf{F}_{self} = \mathbf{0}$. We have here a contradiction following from Frisch's assumptions (1)–(4); thus (1)–(4), as Frisch interprets them, are inconsistent.[7]

So Frisch's 'core assumptions' really are inconsistent, since a contradiction can be derived from them. The obvious point of contention is his construal of the LFE, as he himself notes:

[T]he inconsistency is most plausibly seen as arising from the fact that the Lorentz force equation of motion ignores any effect that the self field of a charge has on its motion. The standard scheme treats charged particles as sources of fields and as being affected by fields—yet not by the total field, which includes a contribution from the charge itself, but only by the field external to the charge. (Frisch 2005a: 35)

However, he has his reasons for taking his version of the LFE as canonical. The question, of course, is not whether the inconsistent assumptions *are* what Frisch calls 'the standard scheme', or whether they *are* the 'fundamental principles' or 'core assumptions' of CED. Even if many people would be uncomfortable with such terminology, that shouldn't distract us from the more fundamental question here: is there something interesting or important that we can learn as philosophers of science—for example something to learn about how science works that we weren't previously aware of—that follows from the fact that the particular set of propositions Frisch *has* picked out are inconsistent?

4.4 Defending Frisch

The community has been quick to criticize Frisch's claims, with Muller (2007) and Belot (2007) leading the way (all references to Muller and Belot will refer to these papers). But if we employ the theory eliminativism method, one sees that the criticisms are ultimately not successful.

---

[7] This isn't quite how Frisch presents the inconsistency, although it is closely related. For a reconstruction of Frisch's presentation, see my original paper (Vickers 2008).

4.4.1 *Muller*

A weighty dependency on the concept *theory* is obvious in Muller's paper. In §1 he writes, 'we define the theory of CED in order to know exactly what is the object of Frisch's provocative charges.' He then goes on to give a *different* construal of the theory to that given by Frisch, most obviously because he introduces a different LFE. So in fact what Muller presents is *not* the 'object of Frisch's provocative charges', and he gives the impartial reader no reason to accept this version of the theory as *the* theory over Frisch's version.

With this as the starting point Muller ought to argue that Frisch's inconsistency proof goes through fine, but that the 'thing', the 'unit of analysis' he finds to be inconsistent is not the theory of CED.[8] But Muller apparently doesn't notice this difference between two different possible interpretations of the content of the theory. Instead he writes that Frisch 'has applied CED inconsistently', and that 'the logic of the proof that has led him to this conclusion [is] flawed' (§1). He goes on to say that Frisch makes 'two contradictory assumptions' such that 'we already have a contradiction by ∧-introduction!' (2007: 261).

But this paints Frisch in a far more confused light than he ought to be. Muller seems to see *Frisch himself* as inconsistent, meaning that what he takes the theory to be is not held constant. But it is perfectly possible to interpret Frisch as he explicitly intends to be interpreted, as *consistently* (unwaveringly) employing the LFE $\mathbf{F}_{tot} = q(\mathbf{E}_{ext} + \mathbf{v} \times \mathbf{B}_{ext})$, and thereby presenting an inconsistent set of propositions.

So here we have a striking example of the potential miscommunication that can arise when one theory-name is used to refer to two subtly different units of analysis. And, in addition, the sort of argument that can arise when it is assumed that there is a single set of propositions (models, etc.) which just *are* ('is' of identity) the content of the theory. As I said in Chapter 3, I see no reason why we should expect theories to have an exact 'content' in this way. And if they don't, this provides some ontological motivation for theory eliminativism. But again, I don't need to make this argument, since pragmatic motivation is sufficient: we can debate everything we want to debate about inconsistency in science without making any assumptions about

---

[8] Frisch (2008) makes this same point independently of my own discussion in Vickers (2008).

'theories'. In the context of this chapter this is done by taking Frisch's propositions (1)–(4) and examining whether there is something interesting and/or important about the fact that those particular propositions are inconsistent. Caught up in its particular preconceptions about *theory*, and about *classical electrodynamics* in particular, Muller's paper falls short of showing that there is no interest in finding Frisch's propositions (1)–(4) inconsistent.

4.4.2 *Belot*

Belot makes a concerted effort to be more sensitive to Frisch's position. He tells us that the LFE comes in two different versions, one of which (Frisch's) is $\mathbf{F}_{tot} = q(\mathbf{E}_{ext} + \mathbf{v} \times \mathbf{B}_{ext})$. He presents the contradiction just as Frisch does, in terms of energy radiated, as $E_{rad} = 0$ & $\sim E_{rad} = 0$ (2007: 271, fn. 11). The other version of the LFE Belot writes as $\mathbf{F}_{EM} = q(\mathbf{E} + \mathbf{v} \times \mathbf{B})$, and he soon makes it clear that this represents $\mathbf{F}_{tot} = q(\mathbf{E}_{tot} + \mathbf{v} \times \mathbf{B}_{tot})$. The purely self-field part of his LFE can then be written as $\mathbf{F}_{self} = q(\mathbf{E}_{self} + \mathbf{v} \times \mathbf{B}_{self})$.

As already noted in §4.2, stating the self-force in terms of self-fields is basically uninformative unless one provides a model of the particle in question. This Belot does, by introducing a rigid, spherical charged body with 'continuum many parts' (2007: 269). The self-field is then taken into account by considering the force felt by each infinitesimal part of the rigid body by the other parts. Since these 'self-forces' are just special cases of external forces, Belot's self-LFE (and thus his total-LFE) takes the same form as the external-LFE. Now, as also noted in §4.2, bringing in a particle model like this brings to the surface many questions about the nature of the particle which are, at best, very difficult to answer.[9] But, assuming these questions can be answered *somehow*, one can attempt to show that the Frisch inconsistency is overcome as follows.

Consider an accelerating charged particle within an imaginary sphere. From the discussion in §4.2 we know that for energy to be conserved we need the energy flowing over the surface of the sphere in a given time to

---

[9] First of all, Belot's model is an obvious idealization given the conflict between its rigidity and special relativity. But when one attempts to de-idealize, a rather different inconsistency raises its head here. See §7.4, this volume, for discussion.

equal the change in field energy within the surface in that time plus the change in particle energy within that time. In other words,

$$E_{\text{over surface}} = \Delta E_{\text{field}} + \Delta E_{\text{particle}} \quad (4.1)$$

Is this story reflected in the formalism?

First, working with the MEs one can take the surface integral of the Poynting vector to get the energy flow over the surface per unit time:

$$\frac{1}{\mu_0} \int_S \mathbf{E} \times \mathbf{B}.d\mathbf{S}$$

(since **E** and **B** come from the MEs, they are the total fields). Now the MEs can be used to establish the following equality:

$$\frac{1}{\mu_0} \int_S \mathbf{E} \times \mathbf{B}.d\mathbf{S} = -\frac{d}{dt} \int_V \frac{\mathbf{E}.\mathbf{E} + \mathbf{B}.\mathbf{B}}{8\pi} dV - \int_V \mathbf{J}.\mathbf{E}_{tot} dV$$

The first term on the right hand side is the change in field energy inside S per unit time. So, for energy to be conserved, we want the second term on the right hand side to be the change in particle energy per unit time (cf. equation (4.1)).

Since we are dealing with a single particle instead of a current or charge distribution, we exchange the **J** in the final term for $q\mathbf{v}$ and do away with the volume integral. Also, we can multiply all terms by a tiny increment of time dt; this will allow us to talk of actual amounts of energy instead of energy per unit time. This then gives us,

$$\frac{dt}{\mu_0} \int_S \mathbf{E} \times \mathbf{B}.d\mathbf{S} = -d \int_V \frac{\mathbf{E}.\mathbf{E} + \mathbf{B}.\mathbf{B}}{8\pi} dV - q\mathbf{v}.\mathbf{E}_{tot} dt$$

Now the equality says that the energy flowing across the surface in time dt is equal to the change in field energy in that time minus $q\mathbf{v}.\mathbf{E}_{tot}dt$. So if the latter expression gives us the change in particle energy, we are on to a winner.

We get the change in particle energy from considering the LFE, and using the fact that the work done on the particle in time dt is equal to the change of kinetic energy of the particle in that time. Now the work done equals force times distance: $W = \mathbf{F}.d\mathbf{l}$ for a tiny distance $d\mathbf{l}$. And since we can write $\mathbf{v} = d\mathbf{l}/dt$ we can also write the work done as $\mathbf{F}.\mathbf{v}dt$. Applying this to the LFE we get,

$$\mathbf{F} \cdot \mathbf{v}dt = q\mathbf{v} \cdot \mathbf{E}dt + q\mathbf{v} \cdot (\mathbf{v} \times \mathbf{B})dt = q\mathbf{v} \cdot \mathbf{E}dt$$

since $\mathbf{v}.(\mathbf{v} \times \mathbf{B}) = 0$. So the change in energy of the particle in time dt is given by $q\mathbf{v}.\mathbf{E}dt$. So if the $\mathbf{E}$ in the LFE represents the *total* field $\mathbf{E}_{tot}$, as it does in Belot's LFE, then energy is conserved. And of course if the $\mathbf{E}$ represents the external field only, $\mathbf{E}_{ext}$, as it does in Frisch's LFE, then energy is not conserved. Assuming energy *is* conserved (and assuming the legitimacy of the given derivation), it is clear that Belot's LFE (but not Frisch's LFE) delivers a consistent picture.[10]

Having recognized that his and Frisch's construals of the LFE are different, Belot now asks what might initially be thought of as the crucial question: Which version of the LFE 'really belongs to the theory'? He then gives voice to some of the *prima facie* advantages of his version over Frisch's:

(i) '[Frisch's CED] does not deserve to be called a theory precisely because it is inconsistent' (2007: 277);
(ii) the total-LFE 'is fundamental', while the external-LFE 'is naturally seen as arising in the course of taking a useful approximation' (2007: 272);
(iii) '[T]he total-field version of the law should be preferred, as taking into account all of the actors involved' (2007: 272);
(iv) '[E]nergy is not conserved in this [Frisch's] theory' (2007: 272);
(v) if you follow Frisch 'you can show that just about any theory is inconsistent' (2007: 275f.).

These objections initially appear to amount to a strong case against Frisch, and there are those in the community who have considered this the end of the matter. But, as in Muller's paper, we see a strong dependence on the notion of a 'theory', and in particular the idea that there is a fact of the

---

[10] In personal communication Frisch has suggested to me that one cannot make the move from J to $q\mathbf{v}$ without considering the nature of the charged particle. However, this is a standard move in the relevant literature, and I don't think Frisch means to claim that Belot's LFE is *also* inconsistent with the MEs, etc. as his own LFE is. The important point is that any attempt to actually *make use* of Belot's LFE causes all sorts of problems, whereas the Frisch LFE is extremely useful (see below).

matter about which version of the LFE 'really belongs to the theory'. This line of reasoning completely leaves aside the more important question of whether Frisch has nevertheless introduced an interesting and/or important inconsistency. This will be demonstrated as I consider each of Belot's criticisms in turn.

(i) This statement—'Frisch's CED does not deserve to be called a theory precisely because it is inconsistent'—shows up very clearly the danger of putting too much emphasis on the concept *theory*. Belot has a concept of theory such that a theory is consistent by definition. Frisch obviously does not. With these two different conceptions in place it is very easy for arguments and disagreements to materialize and perpetuate. Frisch says that his version of CED is a 'theory' and Belot says that it is not. But once one realizes that they are using the word 'theory' in two different ways, this apparent disagreement dissolves. There is no more disagreement here than if Frisch were to say that the symbol 'L' is a letter (in the sense of a member of the alphabet) and Belot were to say that the symbol 'L' is not a letter (in the sense of a written message). Clearly what is distinctive about the case of 'theory' is that the difference between the two meanings is much more subtle, and not easily decided by context. In addition, in the case of the word 'letter' there are two well-defined, socially embedded meanings, whereas in the case of 'theory' things are not so clear cut.

How, then, should we understand Belot's claim that Frisch's version of CED cannot be a theory precisely because it is inconsistent? Or, to put it as Belot really intends it, how is it that Frisch's CED cannot be a theory precisely because it is *known* to be inconsistent?[11] The statement makes sense if Belot is engaged in an investigation into propositions which can be accepted as candidates for the explanatory truth of the relevant domain of phenomena. Of course Belot realizes that CED is superseded, and therefore is no longer a candidate for the explanatory truth of EM phenomena. But the theory was originally put forward as a candidate for the explanatory truth, and Belot might claim that he is merely considering it as it was, in its proper historical context. And of course, the relevant propositions could not possibly both be candidates for the truth and also known to be inconsistent.

---

[11] This must be what Belot means, otherwise it is too radical a proposal.

But, as I've already stressed, there is more than one way in which propositions can be 'pointedly group', such that the inconsistency of those propositions is interesting or important. Frisch is clearly interested in a different grouping of propositions to Belot. He writes,

> [I]n accepting a theory, my commitment is only that the theory allows us to construct successful models of the phenomena in its domain, where part of what it is to be successful is that it represents the phenomena at issue to whatever degree of accuracy is appropriate in the case at issue. (Frisch 2005a: 42)

No mention here of 'the truth'; instead Frisch is concerned with constructing successful models of the phenomena. It might still sound like there is 'one theory' with 'one content', and that Frisch merely has a different take on how we should 'accept' or 'commit to' that theory. However, in other passages it becomes clear that the very content of Frisch's CED is defined by the commitment in question. He writes,

> [CED is] the most common theoretical approach to *modelling* the interactions between charged particles and electromagnetic fields. (Frisch 2005a: 1, my emphasis)

> Throughout my discussion I will refer to the scheme used to *model* classical particle-field phenomena as a 'theory'. (Frisch 2005a: 26, my emphasis)

In other words, whereas Belot is focused on propositions which were considered candidates for the truth, Frisch is focused on propositions which were (and still are) used to construct successful models of the phenomena. But if some models require some propositions, and other models require contrary propositions, why shouldn't Frisch's set of propositions be inconsistent? And, regardless of whether we call Frisch's assumptions (1)–(4) 'the theory of CED' or not, why shouldn't that be an inconsistency which carries important lessons for how science works?

(ii) Given the discussion in Chapter 3, Belot's point here is quickly discerned. When he writes 'The total-LFE is fundamental' he is saying that it is the LFE in *this* form that scientists working at the time would have put forward as a serious candidate for the explanatory truth.[12] And when he writes 'the external-LFE is naturally seen as arising in the course of

---

[12] On p.264 Belot describes CED as a 'less than fundamental theory'. This is not a problem: at this point he is talking about CED from *our* perspective, whereas when he describes the total-LFE as 'fundamental' he is thinking about the perspective of scientists working at the time.

taking a useful approximation' he is making the point that Frisch's external LFE was always believed to be false, since scientists working at the time knew that accelerating charged particles *did* experience a force due to self-radiation.

However, Belot seems to do little more here than state that this means that it is the *total* LFE which belongs to the theory, so that the theory is consistent after all. Once again this talk of 'the proper content of the theory' is unhelpful. Adopting theory eliminativism things look as follows. Belot acknowledges that it would be interesting/important if the propositions believed to be serious candidates for the explanatory truth were inconsistent (like the Pauli inconsistency in Chapter 3). He is keen to note that these propositions are *not* inconsistent. But this does nothing to refute the claim that the inconsistency of the propositions Frisch has presented is an interesting/important inconsistency. There may be an implicit claim in Belot's argument that inconsistency could *only* be interesting if one is focused on fundamental assumptions, but he neither states nor presents an argument for this. He does recognize that scientists regularly *wrote down* the equation in Frisch's form, so its historical relevance, at least, is not in doubt.

(iii) What does Belot mean by the LFE taking into account 'all of the actors involved'? At first this seems pretty clear cut. Whether we work with Frisch's or Belot's version of the theory, the MEs are included, and they tell us that accelerating charged particles radiate energy. By energy conservation (also in both interpretations of 'the theory') this will cause the particles to lose energy, and from this loss of energy we can infer a force on the particle. Therefore, it would seem, the self-fields should play a part in the LFE; they are 'involved'.

The focus here is the 'relevant domain of phenomena', which plays a part whether we are interested in propositions which are candidates for the explanatory truth (Belot) or propositions which are used for model-building (Frisch). In the terminology of theory eliminativism Belot's claim can be translated as follows: Frisch's inconsistency will not be significant, because the propositions in question are not 'pointedly grouped', because they do not properly correspond to the relevant domain of phenomena. The point is that sets of propositions in science are often put together according to how they account for a well-defined domain of phenomena. The natural boundaries of the domain of phenomena create natural boundaries in the set of propositions put together to explain those phenomena. But, according to Belot, Frisch's propositions leave out one of the key ingredients in the domain of EM phenomena: self-force effects. Thus any inconsistency in Frisch's propositions will carry little significance for how science works, because the propositions do not properly belong together as a 'pointedly grouped' set.

However, if this represents Belot's argument, then he seems to go against it in another passage, where he apparently agrees that there are no relevant phenomena which require the total-LFE for their explanation. Drawing on Jackson he tells us that the external-LFE is 'empirically adequate down to the level at which quantum effects begin to appear' (Belot 2007: 274). Frisch tells us the same thing: quantum effects become important before self-radiation effects do (Frisch 2005a: 42). Now, everyone agrees that CED need not concern itself with quantum effects: these are insignificant in the relevant phenomenological domain. But if self-force effects are equally negligible, then we must ask why these *must* feature in the explanandum domain of phenomena. There is clearly a significant sense in which they are *not* 'involved', just as quantum effects are not.

If he accepts that Frisch is focused on a domain of phenomena which doesn't include the self-force, Belot may accept Frisch's version of the LFE. But then he might still object to the inconsistency, saying that if the self-fields are to be left out of the LFE they ought *also* to be left out of the MEs. Self-fields should be rejected altogether since they are apparently insignificant in the domain. However, of course it is only the *effects of self-fields on their host particle* which are insignificant. The self-fields themselves are perfectly significant, and were being readily 'observed' and manipulated in experiments as early as the 1890s, X-rays being the first important manifestation. So although self-*forces* are negligible within the relevant domain, self-*fields* are not.

Therefore if we interpret Belot's 'being involved' as 'being significant' we see a possible motivation for grouping together Frisch's inconsistent set of propositions. Frisch might argue as follows. Self-*radiation* comes under the effects of particles on fields, and therefore is the business of the MEs. This is non-negligible (in non-quantum contexts), and thus the MEs ought to include it. By contrast the self-*force* comes under the effects of fields on particles, and therefore is the business of the LFE. This *is* negligible, and thus the LFE ought *not* to include it (just as it doesn't include quantum considerations). A consistent account of particle-field interactions can't accompany such an emphasis on significance. That this underlies Frisch's account is suggested (Frisch 2005a: 16) when he draws on Rohrlich and Harding. They cite electrodynamics as an example of an 'established theory', which is mature and successful and 'a permanent part of science', but which has been superseded (Rohrlich and Hardin 1983: 603). The boundary of such a theory is defined according to its validity limits, which could be interpreted as placing the self-force outside the theory and the self-field inside. In short, if one looks to the domain of classical EM phenomena to be explained (that is, EM phenomena observable within classical limits), then Frisch's inconsistent propositions seem to group together very nicely as the most natural explanans to explain all and only those phenomena.

Belot may still object that self-forces are clearly 'involved', because it follows from the rest of the theory that they are. In other words, even if the effects of self-forces are not significant in the relevant domain, one can still infer that these effects exist by making inferences from the MEs. But this is far from straight forward, and Frisch can certainly resist. *Prima facie* there are two different, overlapping and closely related, but both 'pointedly grouped', sets of propositions here. One (Frisch's) is grouped together to account for *only* the relevant observable phenomena, and the other (Belot's) is grouped together to account for the relevant observable phenomena plus unobservable phenomena which can be deductively inferred from assumptions necessary to account for the observable phenomena. It certainly isn't obvious that there is *no* good reason to group together propositions according to Frisch's rationale. And in particular, it is certainly not obvious that the inconsistency of Frisch's propositions, so grouped, cannot be interesting or important.

(iv) That energy is not conserved in Frisch's account is stated by Belot on several occasions. However, Frisch has included energy conservation in his set of 'fundamental principles' of the theory, (1)–(4). It appears to me that all Belot is really saying here is that energy conservation is inconsistent with the other propositions in Frisch's theory, and that this just won't do. So this is really just another way of saying that Frisch's CED is inconsistent, and 'something's gotta give'. But of course energy conservation is no more violated in Frisch's scheme than are the MEs (2), or his LFE (3), or even the ontological claims (1).[13] By Frisch's lights, since all of the propositions are used in different circumstances, all of them deserve a place in the theory. But we don't use them all together, at the same time; as Frisch writes,

[F]or a given system we use only a proper subset of the theory's equations to model its behavior, where the choice of equations *depends on what aspect of the interaction between charges and fields we are interested in*. (Frisch 2005a: 40, original emphasis)

Here, once again, Belot is biased by his underlying conceptions of *what theories are* and *how they are used*. But he hasn't told us what is wrong with Frisch's conception, or why *given* Frisch's conception the inconsistency of CED cannot be interesting or important.

(v) According to Belot, Frisch will find that 'any reasonably complex physical theory is inconsistent' (2007: 275). This is because, as Belot sees it, Frisch is pursuing 'a desire to be faithful to the practice of physicists' (2007: 273). And, when we look at that practice, different parts of theories are ignored at different times for the sake of simplification and approximation. According to Belot, Frisch will then include in his theory a statement and its negation, since the negation will represent theoretical contexts where that particular theory-element is ignored.

Once again, the way Belot puts things means we are in danger of disagreeing only about the way the word 'theory' is used. But how we decide to use the word 'theory' is neither here nor there insofar as understanding science goes. Once we employ theory eliminativism Frisch is better described as considering propositions which are chosen depending on how models of the relevant phenomena are constructed. It may well be interesting when such a set of propositions are inconsistent, and it's far from clear that Belot's concern—that many such sets of propositions will be found inconsistent—has any force.

But I'm not convinced that many such sets of propositions *will* be found inconsistent, because Belot has missed an important motivation for Frisch's

---

[13] See Frisch (2005a: 51ff.), for a discussion of how different (unappealing) ontological commitments render the theory consistent.

content selection already noted in section (iii). Frisch can argue that CED is inconsistent because the self-force is *always insignificant in the relevant domain*, even though the self-fields are not. Belot seems to be suggesting that for Frisch a theory will include a categorical statement (such as $\mathbf{F}_{self} = \mathbf{0}$) just because some theoretical feature is only *sometimes* ignored. Frisch can reply that such a feature ($\mathbf{F}_{self}$) must *always* be ignored for the statement ($\mathbf{F}_{self} = \mathbf{0}$) to be included. This given, there is good reason to suppose that even on Frisch's understanding of 'theory' there would be very few inconsistent theories. The inconsistency of CED (for Frisch) is motivated by the peculiar fact that self-fields are significant in the (observable) domain of CED, but self-forces are not. This seems to be fairly idiosyncratic: something that probably wouldn't arise with much frequency in science.

So Belot's criticisms of Frisch's position don't quite hit the mark, or at least don't give the requisite detail to do the intended damage. The recurring oversight on Belot's part is that Frisch's overall philosophy of science is not taken into account; his position is not assessed on its own terms. The question which needs to be asked is not whether Frisch's version of the LFE should be included in the theory, but whether the inconsistency Frisch identifies is interesting and significant given the character of the set of propositions he has selected for analysis. This will be considered in the next section, §4.5.

Before that, I want to respond here to certain critics who may still be less than convinced of the motivation for theory eliminativism, who feel that Belot's own version of the LFE is so well motivated—compared with Frisch's, at least—that there is *much* more justification for calling his unit of analysis 'the theory'. In other words, I wish to provide some further evidence here that—in addition to pragmatic motivation—there is good *ontological* motivation for theory eliminativism. It turns out that, even with a theory as apparently well-defined as CED, serious difficulties arise when we try to pin down 'the content of the theory'. Just one small example of this is the justification for Belot's own version of the LFE, even in the light of his own understanding of 'the theory'. There are important lessons here concerning the different ways in which propositions can be, or fail to be, 'pointedly grouped'. It turns out that the total-LFE is (B1) not useful, (B2) not significant, and (B3) doesn't follow from the rest of the 'theory' (as Belot interprets it).

(B1) As already noted (in §4.2), the self-field part of Belot's LFE—$\mathbf{F}_{self} = q(\mathbf{E}_{self} + \mathbf{v} \times \mathbf{B}_{self})$—requires a model of a charged particle if it is going to be useful, and Belot's paper attests to this when he introduces such a model. Upon deriving a similar equation Jackson remarks, 'To calculate the self-force... it is necessary to have a model of the charged-particle' (1999: 751). Where does such a model come from? Should it be included in Belot's 'theory' given his particular 'point of grouping'?

Recall that Belot's propositions are those which were considered candidates for the explanatory truth of the relevant domain of phenomena. But Belot is quite clear that any model of a charged particle one introduces is not going to be a *serious* candidate for the truth, since it will inevitably be extremely speculative. He writes,

[T]he Maxwell-Lorentz equations no more pick out a structure for microscopic charged particles than Newton's law of gravity picks out a structure for microscopic massive particles: one is free to stipulate a notion of particle for a given investigation. (Belot 2007: 266)

But if particle models are too speculative to put in the theory, then the self-LFE is left as a useless assumption because it cannot function *without* such a model. So it might be suggested that Belot should leave it out altogether, unless he is willing to define 'theory' in such a way that particle models can be included in the theory as well.

(B2) As discussed above in section (iii), even when we add a particle model to Belot's LFE, the latter still isn't significant enough to be of interest in the domain of classical EM phenomena. So, starting any problem of CED with the two halves of Belot's LFE—$\mathbf{F}_{ext} = q(\mathbf{E}_{ext} + \mathbf{v} \times \mathbf{B}_{ext})$ and $\mathbf{F}_{self} = q(\mathbf{E}_{self} + \mathbf{v} \times \mathbf{B}_{self})$—one would immediately ignore the second half and work exclusively with the first. Worse yet, for many problems one would not just ignore the second half, but would actually *change* it to $\mathbf{F}_{self} = 0$ to match Frisch's version of events (see §4.5, this volume). The suspicion arises that there is no warrant to include Belot's $\mathbf{F}_{self} = q(\mathbf{E}_{self} + \mathbf{v} \times \mathbf{B}_{self})$ *except* to ensure consistency. But then it is not included in his 'theory' for the right reasons: since it doesn't play a part in relevant explanations it cannot be a candidate for the explanatory truth.

(B3) As already noted, for Belot the elements of a theory seem to be what were considered serious candidates for the explanatory truth of the relevant domain of phenomena by the relevant community of scientists.[14] Now despite (B1) and (B2), Belot might argue that his self-LFE should stand, because if one believes a set of propositions are true, and then uses truth-preserving inferences to reach another proposition, one should believe that proposition to also be true. However, it might be objected that these considerations do not put Belot's LFE in the theory either.

Recall from the beginning of this section that Belot's LFE was shown to be consistent with the MEs and energy conservation. What was required was for the following equality to hold:

$$\mathbf{F}_{tot} \cdot \mathbf{v} dt = q\mathbf{v} \cdot \mathbf{E}_{tot} dt$$

Belot's LFE succeeded here where Frisch's failed. Now we don't want to meddle with the purely external part of the LFE (it surely belongs in every set of 'pointedly grouped propositions' relevant to EM phenomena, since it is significant, useful, well-confirmed in the domain, etc.), so what is important here is the following equality:

$$\mathbf{F}_{self} \cdot \mathbf{v} dt = q\mathbf{v} \cdot \mathbf{E}_{self} dt$$

This actually gives us quite a bit of freedom vis-à-vis the self-field part of the LFE. Instead of Belot's $\mathbf{F}_{self} = q(\mathbf{E}_{self} + \mathbf{v} \times \mathbf{B}_{self})$ we could just have $\mathbf{F}_{self} = q(\mathbf{E}_{self})$, and claim that the particle only experiences a force due to *electric* self-fields. More generally, the following equation satisfies the equality, where g stands for *any* 3-vector to 3-vector function:

$$\mathbf{F}_{self} = q\mathbf{E}_{self} + \mathbf{v} \times g(\mathbf{B}_{self}) \qquad (4.2)$$

It doesn't matter what vector we cross $\mathbf{v}$ with in the final term: $\mathbf{v}.(\mathbf{v} \times \mathbf{A}) = 0$ for all $\mathbf{A}$.

In fact if we are working with the relativistic version of CED Belot's self-LFE *would* follow from the rest of the theory. This is because the only way to make equation (4.2) covariant is to set $g(\mathbf{B}_{self}) = q\mathbf{B}_{self}$.[15] This would provide us with the necessary warrant for including Belot's total LFE in the

---

[14] Belot is never explicit about this, but it seems to explain several of his comments and some of his objections to Frisch. For more evidence that this is his view, see Belot (2007: 280–1).

[15] To see this, consider a moving charged particle from the perspective of two different frames of reference, relativistically transform F, E, B, and v, and consider what form g must take for equation (4.2) to be covariant.

relevant assumption set (although the equation still wouldn't do any theoretical work for him). However, Belot apparently is not considering the relativistic version of the theory. He writes,

[M]y present concern is with the question whether (1)–(5) are consistent—and special relativistic considerations form no part of (1)–(5). (Belot 2007: 266)[16]

However, Belot's reason to keep things non-relativistic doesn't seem overly motivating. He wants to introduce rigid bodies, which are inconsistent with special relativity, but he has already said that electron models are not part of the theory on his account. At least with relativity on board his total-LFE would be motivated.

In a non-relativistic setting, Belot might try to justify his version of the F_{self} equation by arguing that the self-force is just a special case of the external force, as noted above, and that the two equations must therefore take the same form. But this would require particles to be extended, and Belot has said that the theory shouldn't take a stand on whether particles are extended or points (2007: 266f.). This is no small matter, since the external-LFE and the self-LFE treat particles in very different ways: the former refers to an effect on the particle *as a whole* (acting at the point of the centre of mass of the particle), whereas the latter refers to an effect on different *parts* of a particle. And in addition any justification of Belot's total-LFE would have to include an argument as to why each part of a particle does not affect *itself* (Griffiths 1999: 472).

Thus it would seem that, unless he means to refer to *relativistic* CED, the total-LFE should not be included in the theory *even on Belot's own terms*. Or if he does include it he would need to also include a model of an electron, and show us where it actually features in an explanatory role. But Belot is clear both that he isn't considering relativistic CED, and that he doesn't want to include a model of an electron in the theory. With this in mind I would recommend equation (4.2) as the correct formulation of the self-force part of the LFE for his purposes.

Finally, not only does Belot have some work to do to justify his version of events, but he freely admits some advantages (were other things equal) of

---

[16] Even if Belot is staying silent on whether the equations should be interpreted relativistically, my suggestion is that his version of the LFE would be better motivated if he did not stay silent, but instead brought in relativity explicitly.

Frisch's conceptual scheme. Belot writes, 'The big advantage of this version [the external-LFE] is that it allows one to work with much simpler equations' (2007: 272) and '[I]t appears that at the level of official doctrine and at the level of problem-solving, the external version of the Lorentz force law is taken as standard by physicists' (2007: 273). He also notes of Feynman, 'Much that he says gives the impression that he identifies classical electrodynamics with [Frisch's version of the theory]' (2007: 274). And on p.273 he also writes, '[I]n Jackson's canonical textbook one finds the Lorentz force law given in the external-field form', referring to Jackson (1962: 191). Belot is keen to point out that both authors, Jackson and Feynman, do 'eventually' note that they have neglected radiation reaction. But it is clear that he does see the appeal of focusing on a set of propositions which are central to 99 per cent of the literature. Clearly the concept of 'historical relevance' is not lost on Belot: Frisch's LFE is commonly found in the relevant substrate of science. The question which remains, then, is what can we infer from the fact that Frisch's set of propositions are inconsistent? And can Frisch's construal of the theory be criticized on *its* own terms, as Belot's can?[17]

4.5 The Significance of the Frisch Inconsistency

We have now seen just how different Frisch's conception of CED is from Belot's. The difference in the construal of the LFE is only the beginning. Crucially, different conceptions of what deserves to be called a part of a theory accompany each LFE. On the one hand Belot looks, at least in part, to the doxastic commitments of the relevant community, and which theoretical constituents were considered candidates for the explanatory truth of the relevant domain of phenomena. Frisch, on the other hand, identifies CED with '[T]he most common theoretical approach to modelling the interactions between charged particles and electromagnetic fields' (Frisch 2005a: 1). Since each party insists on calling their unit of analysis 'the theory' or 'CED' there is little wonder miscommunication arises. But apart from this terminological disagreement, there is no reason why both research

---

[17] This chapter departs from my paper (Vickers 2008) in various ways, but especially in the remainder. This is partly because I wish to make different points here, but also partly because I no longer fully agree with what I wrote in that paper.

programmes cannot exist side by side. Dividing the two units of analysis into 'theory$_B$' for Belot, and 'theory$_F$' for Frisch, we can happily say that 'theory$_B$' is consistent and 'theory$_F$' is inconsistent. One might draw here on Kenat (1987: 87) who (himself drawing on a paper by Sylvain Bromberger) distinguishes two types of theory. 'Theories1' are 'theories as techniques for developing answers to problems' (cf. Frisch), and 'theories2' are 'propositions' (cf. Belot).

What is required to do justice to either party is to consider his conceptual scheme taken as a whole, on its own terms. As we saw in the last section, Belot's criticisms fail to do justice to Frisch's underlying conception of *theory*. Frisch at least appreciates that there are two different approaches, and briefly considers the possibility of compatibility. He writes, 'a certain amount of peaceful coexistence between the two rival views of theories is possible, if we realize that they might be talking about different aspects of scientific theorizing' (Frisch 2005a: 12). However, he insists that 'Genuine disagreements exist' (2005a: 11), and CED is meant to represent a situation where Belot's view fails and the models-based view succeeds. His argument is that, on Belot's view, 'accepting an inconsistent theory entails being committed to inconsistent sets of consequences' (2005a: 41). And since CED *is* inconsistent, Belot *is* committed to inconsistent sets of consequences. Thus, 'Inconsistent theories [like CED] may be taken to provide particularly strong support for the importance of "model-based" accounts of theories' (2005a: 12).

However, as we have already seen, Frisch also argues *from* the models-based account of theory acceptance *to* the inconsistency of CED. But he can't have it both ways! At best he has merely established the biconditional:

Frisch's version of theory-acceptance ↔ CED is inconsistent.

If CED is inconsistent we need Frisch's account of theory-acceptance (or something like it). And if we employ Frisch's account of theory-acceptance, then if the content of a theory is defined by what the relevant community *accepts* (as Frisch proposes), CED is inconsistent. So this just testifies to the internal coherence of Frisch's own conceptual scheme, and doesn't impinge on Belot's at all.

Theory eliminativism ensures that we cut through this conceptual mess, and reach the really important questions about science (as opposed to questions about how we talk about science). Frisch *has* presented an inconsistent set of propositions; what of interest can be inferred from the inconsistency he presents? Most obviously we need to ask how Frisch justifies including $\mathbf{F}_{self} = \mathbf{0}$ as a component part of his LFE. Frisch (2005a: ch. 3) has argued at length that there is no fully satisfactory account of self-force effects, despite many decades of research. As noted above, Belot's total-LFE—$\mathbf{F}_{tot} = q(\mathbf{E}_{tot} + \mathbf{v} \times \mathbf{B}_{tot})$—is little use, since the self-force part of this equation—$\mathbf{F}_{self} = q(\mathbf{E}_{self} + \mathbf{v} \times \mathbf{B}_{self})$—is useless without getting into all sorts of difficulties concerning the model of the charged particle in question. But one doesn't *need* to include self-force effects in the LFE to eliminate the Frisch inconsistency. All one needs to do is *take out* any reference to a self-force, by removing the $\mathbf{F}_{self} = \mathbf{0}$ part of Frisch's LFE, leaving behind a 'purely external' LFE: $\mathbf{F}_{ext} = q(\mathbf{E}_{ext} + \mathbf{v} \times \mathbf{B}_{ext})$. It might be objected that then one is completely in the dark as to the self-force, but that is not the case. One can use the MEs and energy conservation to tell us what is happening with any self-force: for example in most contexts one can derive, if one wishes, $\mathbf{F}_{self} \approx \mathbf{0}$ (see, e.g., Jackson 1999: 746f.). Given that this information is already contained within the MEs and energy conservation, Frisch needs to have a story to tell as to why we also need to include $\mathbf{F}_{self} = \mathbf{0}$ as a component part of the LFE. How could we possibly need both $\mathbf{F}_{self} \approx \mathbf{0}$ and $\mathbf{F}_{self} = \mathbf{0}$?

In §4.4.2, where I defended Frisch's position against Belot's criticisms, I suggested a possible motivation for Frisch's LFE: I drew on the 'fact' that the self-force is always insignificant in the domain of CED (the point being to justify Frisch's $\mathbf{F}_{self} = \mathbf{0}$ by appealing to the limits of the relevant domain of phenomena). In section (iii) we saw that Belot and Frisch both draw on Jackson to establish that the external-LFE is 'empirically adequate down to the level at which quantum effects begin to appear' (Belot (2007: 274). The relevant passage in Jackson is as follows: 'Only for phenomena involving such distances [10^{-15}m] or times [10^{-24}s] will we expect radiative effects to play a *crucial* role' (Jackson 1999: 747, original emphasis). However, Jackson doesn't actually mean 'only for' here, for two reasons.

On the one hand the microscopic time and distance noted in the quotation given are derived from the Larmor formula for the radiation power of an accelerating charged particle. But Frisch is keen to stress that the theory he *really* takes himself to be analysing is *relativistic* CED. He writes, 'It [CED] is a *classical* theory only in that it is not a *quantum* theory'

(Frisch 2005a: 29). And when one takes into account the relativistic power radiation formula, one finds that the radiation emitted depends on γ^6 (Jackson 1999: 666). So for a particle travelling at speeds close to the speed of light, the power radiated increases dramatically, and thus the self-force becomes much more significant.

This still leaves us with fairly small times and distances, but there is another factor to take into account: the 'long-term, cumulative effects' (Jackson 1999: 747). For example, since a particle in a synchrotron accelerator is continuously radiating, the effects mount up. In early synchrotrons the loss per turn was only 1000 electronvolts, and yet even this was non-negligible compared to the energy gain of the electron per turn. In more modern synchrotrons, the energy loss is 300 million electronvolts per turn (Jackson 1999: 667f.). And quantum effects can certainly be ignored here: in fact Planck's constant is of the order of 10^{-15} electronvolt seconds! Such considerations are made clear in problem 16.2 of Jackson's book (1999: 769), where an equation is derived for the final orbital radius of an accelerating particle given the initial orbital radius—we have a non-negligible self-force in a non-quantum context.

The consequence for Frisch's account is this: if synchrotron phenomena are to be included in the relevant domain, he can't justify including $\mathbf{F}_{self} = \mathbf{0}$ in the theory on the grounds that the self-force is *always* negligible. One option is for Frisch to restrict the range of 'the relevant domain', to renege on his claim that his conclusions extend to relativistic phenomena such as those associated with synchrotrons. After all, before synchrotron accelerators were built, there really were no known phenomena where the self-force was significant.

But actually Frisch has a much better reply to this problem of non-negligible self-force. Instead of justifying his version of the LFE on the grounds that the self-force is always negligible, he can justify it on the grounds that his version of the LFE is *always used* for phenomena in the relevant domain. Now, we might quickly accept that this will be true in the vast majority of contexts, where it is known in advance that any self-force will be negligible. After all, in such contexts one can justify including $\mathbf{F}_{self} = \mathbf{0}$ on the grounds that $\mathbf{F}_{self} \approx \mathbf{0}$. But the remarkable fact is that Frisch's LFE—$\mathbf{F}_{tot} = q(\mathbf{E}_{ext} + \mathbf{v} \times \mathbf{B}_{ext})$—is used *even when it is known that the self-force is non-negligible.* In other words, one makes use of $\mathbf{F}_{self} = \mathbf{0}$ (as a component part of the LFE) even when one knows full well $\sim(\mathbf{F}_{self} \approx \mathbf{0})$.[18]

Take a typical account of synchrotron radiation, where the self-force is non-negligible. As Frisch (2005a: 33) emphasizes—following Jackson (1962: 578)—the treatment is 'stepwise'. To quote Jackson, '[F]irst the motion of the charged particle in an external field is determined, neglecting the emission of radiation; then the radiation is calculated from the trajectory' (1962: 578). In that first step, one *cannot* leave out $\mathbf{F}_{self} = \mathbf{0}$ and proceed with the purely external LFE, $\mathbf{F}_{ext} = q(\mathbf{E}_{ext} + \mathbf{v} \times \mathbf{B}_{ext})$. This would be fine for determining the force due to external fields, but then consider how we move from here to the particle trajectory. If one has in hand only $\mathbf{F}_{ext}$, then (using Newton's laws) the calculated orbit would be the 'orbit due to external fields'. Now, what scientific and/or historical relevance does this reconstruction of scientific practice have? Surely none, since (i) no practicing scientist talks in this way, and (ii) it's far from clear that this way of talking makes any sense. It doesn't make sense, because although $\mathbf{F}_{ext}$ might be a component part of $\mathbf{F}_{tot}$, and thus enjoy an existence, of sorts, all of its own, the 'orbit due to external fields' can't be a 'component part of the total orbit'. It just doesn't exist in the same way at all, and any sense in referring to it is lost.

This given, we see that Frisch's inconsistent propositions (1)–(4) are most definitely pointedly grouped, and the point of grouping is that they are all used, in precisely the form Frisch presents them, to model phenomena right across the relevant domain. Remarkably, even when we consider qualitatively different phenomena where the self-force is non-negligible, the very same assumptions are used. This goes to show that, when it comes to CED, there is very good scientific sense in making regular use of a set of propositions which are mutually inconsistent.

Any questions concerning the rationality of science and the 'threat' of inconsistency are easily dismissed. First, it is very easy to see that there is no doxastic inconsistency here comparable to the Pauli inconsistency of the previous chapter: $\mathbf{F}_{self} = \mathbf{0}$ was never a candidate for the truth, at least for any

---

[18] There are rare occasions where one tackles such problems in another way. E.g. one can work from the MEs to derive the Abraham–Lorentz equation, which takes into account self-force effects, as Jackson (1999: 747f.) does. But generally things are not done in this way, because (i) it is more difficult, (ii) there is no advantage to doing so, and (iii) there are disadvantages, since the Abraham–Lorentz equation has some very peculiar features. See e.g. Frisch (2005a: 55ff.) for discussion.

context where a particle is accelerating. Instead $\mathbf{F}_{self} = 0$ was used to make progress and reach approximate conclusions, both in contexts where $\mathbf{F}_{self} \approx 0$ and even in contexts where $\sim(\mathbf{F}_{self} \approx 0)$. There *is* an underlying threat of doxastic inconsistency when we turn to the question of the structure of microscopic charged particles, but I leave this discussion for later (§7.4).

What of the 'threat' of ECQ? Scientists definitely were reasoning with a set of inconsistent propositions, and for the most part they used truth-preserving inferences. How did they avoid deriving an arbitrary proposition? Frisch has his own answer to this. He claims that only a subset of the propositions are ever used in model-construction, and that the choice of propositions depends on 'what aspect of the interaction between charges and fields [scientists are] interested in'. However, it seems to me that there are *some* contexts where the whole set of inconsistent assumptions are used in a single piece of reasoning. For example, when one models the trajectories of particles in synchrotron accelerators, one assumes zero self-force in the first step, but then introduces a second step where self-force effects are taken into account (cf. Frisch 2005a: 33). Even if there are two steps, there is an important sense in which this is a single piece of reasoning: one assumption is used in the first step, and results based upon this assumption are used in the second step.

But regardless of how we reconstruct such modelling procedures, there is really no danger of ECQ here. Frisch's own explanation of how reasoning continues—in terms of scientists employing consistent subsets of the inconsistent propositions—is not necessary. We've already seen in the previous chapter that, when scientists know that at least one of their assumptions is definitely false, they proceed more carefully. This is the same for inconsistency as it is for any other type of reasoning with approximations, idealizations, etc. ECQ does not threaten, because explosion can only come in *via* contradictories, and reasoning will be halted as soon as contradictories are derived.[19] Indeed, reasoning will usually be halted long before contradictories are derived, as soon as we reach anything jarring substantially with physical common-sense. At this point a scientist assumes, quite reasonably, that any further inferences will carry zero scientific value.

---

[19] If the calculations were being carried out by a computer, a contradiction or contradictories might not be noticed. But any absurd consequences following therefrom would. Again, scientists are keenly aware that they need to be careful about trusting inferences based on false assumptions, especially when 'dumb' computers are doing the calculations! (Cf. Wilson 2013: 55f.)

4.6 Conclusion

Frisch's construal of the LFE can be split into an external-field part and a self-field part, and the latter states explicitly that, in all contexts, $\mathbf{F}_{self} = \mathbf{0}$. If Frisch is interpreted as making a claim about the 'content of the theory', then it is a bit misleading: everybody working today knows, and everybody working in the relevant scientific communities in the history of science knew, that $\mathbf{F}_{self} \neq \mathbf{0}$. But if we interpret Frisch as presenting pointedly grouped assumptions (ignoring the question of 'what the theory is'), then it is clear that he has a point. We have seen that his version of the LFE can be pointedly grouped with the MEs given that (i) self-forces are always negligible, or nearly always negligible (depending on how the target domain of phenomena is delineated), and (ii) in practice $\mathbf{F}_{self} = \mathbf{0}$ is assumed even in contexts where the self-force is known to be *non*-negligible.

The assumption $\mathbf{F}_{self} \approx \mathbf{0}$, derivable from the MEs and energy conservation, may be sufficient to explain relevant phenomena, but it is too vague to employ usefully in calculations. In addition, in practice scientists don't first assume $\mathbf{F}_{self} \approx \mathbf{0}$ and then use this to justify the use of $\mathbf{F}_{self} = \mathbf{0}$ in a separate step. $\mathbf{F}_{self} = \mathbf{0}$ is there right from the start as a component part of the LFE (as found ubiquitously in the relevant scientific literature). This given, it is easy to see why Frisch (2005a) considers his version of CED to *be* the theory: other versions of the LFE which avoid the Frisch inconsistency—the total LFE and the purely external LFE—have barely any role at all in the relevant science. Eliminating *theory*, Frisch's assumptions (1)–(4) are very pointedly grouped, but the point of grouping these other versions of the LFE with the MEs, energy conservation, etc., is less obvious. One could perhaps insist that it is the purely external LFE which is the real LFE, but that in the course of solving any problem one immediately draws on the fact that $\mathbf{F}_{self} \approx \mathbf{0}$ follows from the MEs to justify changing the LFE into the Frisch LFE as a useful approximation. But this is a reconstruction designed to meet a preconceived philosophical desideratum. One may of course decide from the outset that one wants to ensure consistency, and choose a version of the LFE which fits the bill, but then one is reconstructing science according to one's philosophical biases instead of according to the real history and practice of science.

The presence of $\mathbf{F}_{self} = \mathbf{0}$ and $\mathbf{F}_{self} \neq \mathbf{0}$ in calculations about EM phenomena might make one wonder about the rationality of science, in some sense. But it should not. The question of how scientists restricted their inferences to avoid deriving anything and everything by ECQ is based on the absurd assumption that scientists are deductive machines, uncritical of the consequences of their inferences even when they know that they are reasoning with (at best) only *approximately* true assumptions. But in such circumstances scientists know that there is the danger of truth-preserving inferences leading them astray, and are careful to examine the consequences of any such inferences on conceptual and/or empirical grounds. In the case of CED (as in the case of Bohr's theory) we can understand the sense in making inferences from mutually inconsistent propositions in the same way that we understand the sense in making inferences from approximation and idealization assumptions generally. And quite obviously the Frisch inconsistency is not serious in the way the inconsistency identified by Pauli in the later Old Quantum Theory (OQT) was serious. When we turn to the question of how we should 'complete' CED, fully including self-force effects, then the structure of charged particles becomes important. And when we turn to the structure of elementary charged particles there may be a doxastic inconsistency lurking. But the nature of this other inconsistency is sufficiently different from the Frisch inconsistency that it deserves separate discussion (see §7.4, below).

Despite the relative insignificance of the Frisch inconsistency, there have been some important lessons along the way. For example, we learn that the way we delineate domains of phenomena can lead us to work—regularly or even exclusively—with a set of mutually inconsistent propositions. One may speculate that, although in the CED case the inconsistency is extremely obvious, there could be cases where the inconsistency was not noticed, and where the propositions are committed to in a doxastic way. And CED shows us that the propositions best suited to explaining a certain, well-defined, unified domain of phenomena may be inconsistent because inconsistency is preferable to facing up to overwhelming conceptual complications that have no explanatory use, at least in the relevant domain.

Turning to metaphilosophy, most important here are the lessons of just what a mess we can get into if we are not careful with our use of the word 'theory', as Frisch (2005a), Muller (2007), and Belot (2007) are not. Clearly there are a whole range of ways of selecting propositions associated with EM phenomena, many different possible 'points of grouping'. Using the word 'theory' for any such set of propositions is bound to lead to miscommunication, especially when two such assumption sets have significant overlap. In such a case, protagonists can talk past one another because they *think* they are both talking about the same 'thing' ('the theory', that includes 'the MEs' and 'the LFE'), when in fact there are important differences.

I hope to have shown that theory eliminativism goes a long way towards eliminating such difficulties. If Frisch had, from the start, presented his set of propositions (making no reference at all to 'CED' or 'the theory'), and explained why it is interesting and/or important that that particular set of propositions is inconsistent, then Muller and Belot's research time would have been more usefully spent. Although, in fact, if Frisch had taken this route it would have led him to think harder about why/whether the inconsistency mattered to philosophy of science, and he might have decided not to make the claim at all. He now admits, 'I am inclined to agree with my critics that this inconsistency in itself is less telling than my previous discussions may have suggested' (Frisch 2008: 94). Of course, I think this is right: there is no doxastic inconsistency, and no threat of ECQ. Thus perhaps the title of Frisch (2004) suggested greater scientific drama than the reality. But, as I have shown, there is some residual interest in noting how the most natural explanandum domain of phenomena brings us to make regular *use*, at least, of inconsistent propositions. Most dramatically, we have seen that scientists persist in using the inconsistency-causing LFE even when they know full well that it is not even approximately true. This shows just how innocuous inconsistency in science can be.

Another lesson from this case study concerns how the delineation of a domain of phenomena can act as a 'point of grouping' for a set of propositions. Domains can clearly be drawn up in all sorts of ways: in the CED case the two major issues were whether we include self-force effects (given how insignificant these effects are relative to the other phenomena of interest), and whether we include relativistic phenomena. The knock-on effects for the grouped propositions are then whether we make our equations relativistic, whether we introduce 'self-force' into the LFE in some way, whether we include claims about particle structure, and also whether we include idealization assumptions or modelling assumptions which readily feature in successful explanations. Apart from the choices we have here, even once we've made our choice there will be difficult decisions about which propositions to select. If we include something about the self-force à la Belot, do we then have to include something about particle structure so that the equation isn't redundant? And if we include certain modelling assumptions à la Frisch, how do we select the set of models in question? And in both cases there are questions to ask about whether we want to include the (deductive?) closure or partial closure of the propositions, whether we want to put an emphasis on the viewpoint of those working at the time or the modern viewpoint, and whether we want to focus on the commitment of key individuals or certain, selected scientific sub-communities.

The beauty of theory eliminativism is that we just don't have to answer these questions. We are no longer in the business of selecting groups of propositions just for the sake of it and giving them names like 'CED'. Instead one asks certain questions about how science works, and if one draws on a given set of propositions one explains why it is appropriate to introduce that particular set of propositions in the context of answering the particular question at hand. There are no sets of propositions one can draw on 'for free' anymore, just because *that's the theory*. Further, one doesn't have the problem of vagueness anymore: however we define 'theory' the content of a specific theory is going to be vague. This will often lead to difficulties when we come to ask questions such as 'is the theory inconsistent or not?' But with theory eliminativism any selected set of propositions either will or won't be inconsistent. If one can justify why one has selected that particular set of propositions—for example by showing that the resultant inconsistency teaches us something interesting about how science works—then any problem of vagueness disappears.[20]

One might be convinced about the difficulties that arise with the concept *theory*, but want to go pluralist instead of eliminativist. In §4.5 I introduced two possible meanings for 'theory', as introduced by Kenat (1987): 'Theories1' and 'Theories2'. Now one might admit, given all of the complications I've introduced, that we need more than these two disambiguations, but insist that nevertheless we can stick with a pluralism of theory-concepts. Eventually different terms might be introduced, just as the term 'jade' split

---

[20] If one defines a 'point of grouping' for one's assumption set, then the *full* set will be vague. But usually it will be enough to focus one's attention just on assumptions which definitely *are* in the set, given the 'point of grouping'. E.g. if assumptions definitely in the set are inconsistent then *however* one completes it the full set will remain inconsistent.

into the terms 'jadeite' and 'nephrite'.[21] However, *theory* is not like *jade*. We used to think that jade was a natural kind, and now we think that jadeite and nephrite are natural kinds. But *theory* is a much more abstract concept, and what it should be taken to mean depends much more on how it can be analytically useful than on what it corresponds to in the world. And, as the plurality of theory-concepts grows (recall Chapter 2 here), pluralism makes less sense. If we get to the stage where we have to explain what we mean by 'theory7', and justify why we have used that concept over the other possibilities, then we might as well have eliminated theory-talk altogether in the first place.[22]

Theory eliminativism makes even more sense in any context where scientists and philosophers are not especially comfortable referring to the object of analysis as 'a theory'. Bohr's atom is relevant here, of course, since it has often been referred to as Bohr's *model* of the atom. But our journey takes us even beyond 'models': the importance of inconsistency does not depend on the object of analysis being the kind of thing usually referred to as a 'model' or 'theory'. As we saw in the Pauli inconsistency, one obvious way in which an inconsistency can be important for science is when we have a set of propositions, committed to as our best shot at candidates for the truth, and somehow pointedly grouped. It turns out that the point of grouping does not have to appeal to a well-defined domain of phenomena (as in CED), or even explanatory work (as in OQT). In the next chapter we will see that an individual question can act as the unifier for a set of propositions, the inconsistency of which can be extremely interesting and important, even if few people have ever been inclined to refer to the relevant set of propositions as 'a theory'. Nevertheless, theory eliminativism is still applicable, since philosophers do insist on making use of a theory-name. That theory-name is 'Newtonian cosmology'.

---

[21] See Piccinini and Scott (2006) and the references therein for more on 'concept splitting'.
[22] In Vickers (forthcoming) I investigate these issues in more depth.

5
Newtonian Cosmology

5.1 Introduction

The focus of attention in this chapter will be what is sometimes called (an old version of) 'Newtonian cosmology'. In 1895, Hugo von Seeliger made the remarkable claim that a set of natural (Newtonian) assumptions concerning forces in the universe—assumptions which had been in place since Newton himself, for over two hundred years—are mutually inconsistent (Seeliger 1895). This stimulated much debate over the years and decades which followed, with the latest additions made by Norton (1993, 1995, 1999, 2002) and Malament (1995). However, one of the most obvious questions the philosopher of science might ask has been largely neglected: If there is an inconsistency, as claimed, why was it that it went unnoticed for two hundred years? What is the nature of the inconsistency which made this oversight possible?

Our first thought might be that this example will mirror the Pauli inconsistency in OQT, which was implicit in the commitments of the community for eight years before Pauli finally noticed it. In that case one of the main factors was that the derivation of the contradiction is rather complex: the inconsistency certainly doesn't just jump out at you when you look at the relevant propositions. But one look at Norton's reconstructions of the inconsistency in Newtonian cosmology shows that no such quick answer will be forthcoming here. For one thing, there are several inconsistencies, which (we might think) ought to increase the chance of noticing the problem. But even more remarkably, it would seem that in each case a contradiction follows from a few basic propositions in a few simple steps. The inconsistencies are, as Malament puts it, 'so close to the surface that they are hard to miss' (1995: 489). This in itself seems to contradict the fact that many great scientists *did* miss the inconsistencies for a period of two hundred years! Otherwise we would apparently have to admit either that scientists made a serious commitment to what they knew to be impossible, or that they were blind to some of the most obvious consequences of their beliefs.

This suggests that something has gone amiss in Norton's reconstruction. One may wonder, for example, whether the propositions in question are as historically relevant as has been made out. As we will see, this is part of the story. But in addition there are several complications to work through to understand the inconsistencies properly: they are not as simple and straightforward as Norton's and Malament's papers suggest. Thus, after a brief section (§5.2) in which I introduce 'Newtonian cosmology' and consider the concept of a 'theory' in this context, I turn to the details of the inconsistency claims which have been made. Four different inconsistencies are distinguished, which are grouped into two *types* of inconsistency discussed separately in §§5.3.1 and 5.3.2. This analysis uncovers certain complications in the science, and in the relationship between Norton's reconstruction and the real history of science, which then help us in §5.4 to understand why the inconsistencies went unnoticed for so long. The chapter concludes in §5.5.[1]

5.2 The Concept *Newtonian Cosmology*

What is 'Newtonian cosmology'? Ask any modern day astrophysicist and they will immediately talk about the *modern* theory considered by Malament in his reply to Norton (Malament 1995). But, following Norton (1993, 1995, 1999, 2002), what we're considering here is the *early* theory, the theory as it existed before certain advances in the twentieth century. However, ask any modern day astrophysicist about *this* and it isn't clear he or she would know what you were talking about at all. The concept *Newtonian cosmology*, in this sense, isn't at all well-defined.

This is made clear when one considers the titles of Norton's papers on the subject. At first he refers to 'A Paradox in *Newtonian Cosmology*' (Norton 1993), which he keeps up in his 1995 paper: 'The Force of *Newtonian Cosmology*: Acceleration is Relative' (Norton 1995). However, there is

---

[1] This chapter is a development of Vickers (2009a).

then a shift in emphasis in his 1999 paper, entitled 'The Cosmological Woes of *Newtonian Gravitation Theory*'. And this new reference to 'Newtonian Gravitation Theory' is kept up in the title of his 2002 paper: 'A Paradox in *Newtonian Gravitation Theory* II' (Norton 2002). On the face of it one might then ask 'Are we talking about Newtonian Cosmology or Newtonian Gravitation Theory?' But this wouldn't be fair—Norton knows exactly what he's talking about: a certain set of propositions. It's just that he isn't certain what to call them. He obviously realizes that some of them have a cosmological character in that they concern the properties of the universe as a whole, and some of them have a Newtonian or gravitational character, such as Newton's laws of motion, and Newton's law of gravitation. In this context the benefits of theory eliminativism are obvious enough: one can talk in terms of sets of propositions which are inconsistent in interesting and important ways. What we decide to name these sets of propositions (if anything) is neither here nor there. In this case it is especially clear that the motivation for ignoring questions about the 'content of the theory' is not just pragmatic, but ontological.

The point of grouping the propositions together in this case is how they come together to answer a single question. We will consider how scientists active between the years 1700 and 1900 would have answered the following:

(Q) What is the net gravitational force on a given test particle at an arbitrary place in the universe?

When this question is asked certain propositions are naturally drawn together. Norton (1995) introduces various such propositions as follows:[2]

(a) Newton's three laws of motion.
(b) Newton's inverse-square law of gravitational attraction.
(b') Poisson's equation with gravitational attraction described in terms of the potential φ.
(c) Matter in the universe is distributed homogeneously (when viewed on a large enough scale) in an infinite Euclidean space.
(d) There is a unique gravitational force on a test mass.
(d') The gravitational potential φ is homogeneous.

---

[2] See Norton (1995: 513–14). In labelling the assumptions, I follow Norton's lead for ease of cross-reference.

It turns out that the question (Q) can be answered in different, contradictory ways depending on which of the given propositions are emphasized. There are essentially four different methods of reasoning which can be employed, which will be introduced in the forthcoming analysis in the following order:

(1) use Newton's law of gravitation;
(2) use Poisson's equation;
(3) use symmetry considerations;
(4) use the gravitational potential.

So in the end the analysis will continue by reference to the question (Q), different methods by which that question would have been answered between the years 1700 and 1900, and the propositions these different methods draw upon. The terms 'Newtonian cosmology' and 'Newtonian gravitation theory' can be used, but as in the previous chapters such names will not be used where it matters. The goal will be to identify the sets of propositions which are inconsistent, and investigate the character and historical relevance of those propositions to see what, if anything, we learn from the inconsistency.[3]

5.3 How was Newtonian Cosmology Inconsistent?

When the noted methods of reasoning provide contradictory answers to our question (Q), one of two principal kinds of contradiction results:

(i) the force on a given test particle is both **F** and **G** where **F**$\neq$**G**, or
(ii) the force on a test particle is both *determinate* (some vector quantity) and *in*determinate (in a sense to be clarified).

These will be tackled in §§5.3.1 and 5.3.2 respectively. The precise inconsistency then depends on which of the given propositions (a)–(d′) the contradiction

---

[3] I have claimed that there is an unfortunate, continuing disconnect between philosophy of science and (i) history of science, and (ii) science itself. I hope to demonstrate this via these case studies. Of course, there are various ways in which I could engage still more with history of science and science itself, especially in this chapter where I am attempting to comment on 200 years (!) of the history of science. For example, when it comes to Newton's laws of motion some of the issues addressed in Earman and Friedman (1973) and related literature might be discussed. I will be content if I have done a better-than-usual job in these respects. Criticisms based on still better historical and scientific work are welcome.

114 UNDERSTANDING INCONSISTENT SCIENCE

is said to follow from. Thus §§5.3.1 and 5.3.2 are each split into further subsections.

5.3.1 *A contradiction of forces*

By 'contradiction of forces' I simply mean that the given propositions can be used to derive the following contradiction:

> (C1) The force on a test mass is **F** and the force on a test mass is **G**, where **F**≠**G**.

This splits into three different claims depending on which of the propositions (a)–(d′) are used to make the derivation.

5.3.1.1 ... *using Newton's law of gravitation?*   Norton's (1993) original paper shows the first possible method of reasoning, and claims that we have a contradiction of forces from propositions (a), (b), and (c):

(a) Newton's three laws of motion;
(b) Newton's law of gravitational attraction;
(c) Matter in the universe is distributed homogeneously (when viewed on a large enough scale) in an infinite Euclidean space.

In greater detail, by (b) we mean:

> (b) The force of gravity $\mathbf{F}_i$ on a test body m_t at $\mathbf{r}$ due to another body m_i at $\mathbf{r}_i$ is given by $\mathbf{F}_i = G \frac{m_t m_i}{|\mathbf{r}_i - \mathbf{r}|^3} (\mathbf{r}_i - \mathbf{r})$.

This gives us the magnitude and direction of the force on our test mass m_t due to *one* other mass m_i. But question (Q) asks what the *net* gravitational force is. By (c), since we are supposing the universe to be infinite and the mass distribution to be homogeneous, there will be an infinite number of masses. Further, (b) comes with no caveat that it doesn't hold beyond a certain distance $|\mathbf{r}_i - \mathbf{r}|$. Thus we must infer that *every* mass in the universe has some effect (however small) on our test mass. Thus there are an infinite number of terms in our sum, and the net gravitational force is represented as the sum-total of all the contributory forces: $\mathbf{F}_{net} = \sum_i \mathbf{F}_i$.[4]

---

[4] For some purposes it is convenient to turn this sum into an integral according to $\sum_i \mathbf{F}_i = \int_V \mathbf{F} dV$, where instead of summing over all (discontinuous) masses we sum over all (continuous) points in Euclidean space, as in Malament (1995: 491). This won't be important here.

At this point, since we are drawing on assumption (c), it is worth pausing to consider what is really meant by the homogeneity of the universe *when viewed on a large enough scale*. Of course nobody ever believed that the universe is *totally* homogeneous, but rather that if you take any arbitrary region of space R of a given large volume V then you will always find the same total amount of mass there (with small deviations from some mean value, which get smaller for volumes of space increasingly larger than V). This actually tells us next to nothing about the density of matter in the vicinity of a given test particle (it could be sat on the surface of a black hole, or be several hundred light years from the nearest massive particle). All we know is the *total* amount of matter in an arbitrary region of space R of volume V, which may include our particle. But this uncertainty in the local matter distribution doesn't transfer to an uncertainty in the force on such a particle, at least insofar as Norton's (1993, 2002) 'lines of force argument' is concerned. All the argument requires is the constant density of matter over different regions R at *some* scale (however big V has to be to achieve this constant density).

Continuing Norton's argument, from the infinite sum we can apparently get different answers depending on how we compute it.[5] If we first consider a spherical region of the universe upon which our particle is sitting—of any given volume V or greater, and situated on any side of our particle—then we get a force **F** towards the centre of that sphere. It can then apparently be shown that the force due to *all other* masses amounts to nothing, since they can be grouped into spherical shells, concentric with the centre of the original spherical region, each of which has no net effect on our particle (see Figure 5.1). Thus our infinite sum turns into $\mathbf{F}_{net} = \mathbf{F} + 0 + 0 + 0 + \ldots = \mathbf{F}$. But the size and direction of the original sphere, and thus the force **F**, was completely arbitrary. Thus Norton claims that the theory is 'logically inconsistent in the traditional strict sense...[because] we can prove within the theory that the force on a test mass is both some nominated **F** and also not **F**, but some other force' (1993: 413; 2002: 186).

In fact no such contradiction can be legitimately derived. Malament (1995) criticizes the reasoning as follows:

---

[5] For the full argument, see Norton (1993 and 2002).

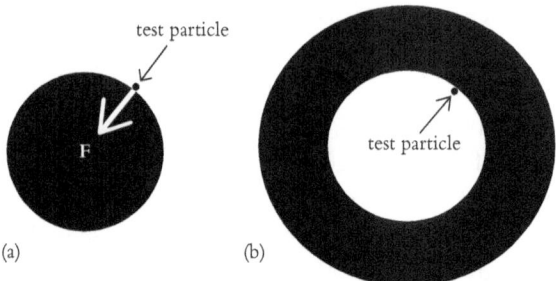

Figure 5.1. A two-dimensional representation of how the forces on the test particle can be added up. Figure 5.1(a) shows the force '**F**' the particle experiences due to the other masses in some large spherical region of the universe, where the location of the test particle is the very boundary of that region. Figure 5.1(b) considers the masses in another large region of the universe, which Norton (1993, 2002) shows will exert zero force on the particle. All other masses in the universe can be taken into account by considering increasingly bigger spherical shells like the one in Figure 5.1(b). The masses in all such shells will exert zero force on the test particle. Thus the net force is, apparently, the original force '**F**' from Figure 5.1(a)

What Norton presents as an argument for inconsistency is better understood as just a vivid demonstration of non-convergence. (A perfect analogue of his argument might be used to 'prove' that, for every integer n, the infinite sum $1 - 1 + 1 - 1 + \ldots$ is equal to n.) ... Rather than asserting that Newtonian theory makes inconsistent determinations of gravitational force ... Norton should have asserted that it makes no determination *at all*. (1995: 491, original emphasis)

In other words, Norton should have noted that not all infinite sums have an answer. For example, mathematicians in the late seventeenth and eighteenth centuries rigorously debated whether the infinite sum $1-1+1-1+\ldots$, known as Grandi's series, is equal to 1, 0, ½, or something else. Today, with the benefit of hindsight, we can look upon these struggles as mere historical curiosities, and say instead that since the series in question is not convergent it has no sum.[6]

The key failure of Norton's argument can be seen by the fact that he groups together the effects of masses in certain regions of space in order to get a result. This is equivalent to bracketing together terms in Grandi's series in order to get a result. But since this bracketing can be done in more than

---

[6] At least, no sum in any physically relevant type of number. See §5.3.2.1 and especially §8.2.4, for some thoughts on how certain issues in the philosophy of mathematics can become relevant.

one way, if this were legitimate you could also show Grandi's series to sum to two different values. The most obvious two are as follows:

$$1 - 1 + 1 - 1 + \ldots = (1-1) + (1-1) + \ldots = 0 + 0 + \ldots = 0$$
$$1 - 1 + 1 - 1 + \ldots = 1 - (1-1) - (1-1) - \ldots = 1 - 0 - 0 - \ldots = 1$$

So does Grandi's series sum to *both* 0 and 1? Since this sort of bracketing is mathematically illegitimate, one must stop at the unbracketed series and conclude that it is equal to *no* value. To reason otherwise is to dabble in inferences which are not truth-preserving.[7]

These facts take on particular significance in the light of Norton's 1999 derivation. He argues (p.274) that the infinite sum can be written as follows:

$$\mathbf{F}_{net} = G\pi\rho\Delta r\hat{\mathbf{x}} - G\pi\rho\Delta r\hat{\mathbf{x}} + G\pi\rho\Delta r\hat{\mathbf{x}} - G\pi\rho\Delta r\hat{\mathbf{x}} + G\pi\rho\Delta r\hat{\mathbf{x}} - \ldots$$
$$= G\pi\rho\Delta r\hat{\mathbf{x}}(1 - 1 + 1 - 1 + 1 - \ldots)$$

where $\hat{\mathbf{x}}$ is a unit vector in *any* nominated direction. This time he considers *hemi*spherical concentric shells, first on one side of the particle—in the (arbitrary) direction of $\hat{\mathbf{x}}$—and then the other—in the direction opposite to $\hat{\mathbf{x}}$—which build up to infinity. But the outstanding question is: why is *this* particular style of summation legitimate and the others not so? Hasn't Norton once again illegitimately grouped together (in hemispheres) the effects of large numbers of masses to achieve his sum? Hasn't he introduced brackets into the reasoning?

This is true to some extent. The real infinite sum, without brackets, is a close approximation to a 3D version of Grandi's series.[8] Strictly speaking Norton shouldn't group together the effects of large groups of masses into hemispherical shells as he does in his 1999 paper. Crucially, however, this time the grouping does not affect the divergence of the summation. The introduction of his brackets is analogous to the following manipulation of Grandi's series:

$$1 - 1 + 1 - 1 + \ldots = (1 - 1 + 1) - (1 - 1 + 1) + (1 - 1 + 1) - \ldots =$$
$$1 - 1 + 1 - 1 + \ldots$$

---

[7] At least this is the normal way to think of these things. Again, one's philosophy of mathematics can matter here.

[8] In fact the real infinite sum could never be presented. Grouping of some sort *has* to be introduced, because in order to consider the universe as homogeneous, we have already grouped together large portions of it.

In other words, the divergence of the sum is preserved by the bracketing. This is the achievement of Norton (1999). Thus, although strictly speaking the infinite sum should *not* be set equal to **k**$(1-1+1-1+\ldots)$—where **k** is the relevant constant vector—the indeterminate nature of the force $\mathbf{F}_{net}$ is preserved by Norton's 1999 analysis, whereas it *isn't* preserved in his 1993 and 2002 analyses. With Norton's 1999 analysis we can be sure that the original series *is* divergent, because the re-ordering and bracketing of a convergent series would never leave us with a non-convergent series.

A further point is in order here. Malament says that what Norton presents is a vivid demonstration of non-convergence. However, more specifically what Norton presents is a vivid demonstration of *alternating* non-convergence. Only when signs in a divergent series alternate is it possible to make quantities cancel out, and achieve various different finite answers through bracketing and re-ordering. Thus there is a sense in which alternating series *don't* diverge, since 'diverge' usually means 'diverge to infinity'. However you bracket and re-order an infinite series which diverges to infinity you get infinity (whether positive or negative). Only with alternating series is it possible to achieve any number of finite answers for the sum of the series. To make this distinction clear, in what follows the word 'indeterminate' will be preferred over 'divergent' to describe the sum of Grandi's series.

With this clarified we can still agree with Malament that propositions (a), (b), and (c) make no determination of the net force. But this isn't because the force is *divergent*, in the sense of 'diverges to infinity'. If it were divergent then we could say that the propositions predict an *infinite* force. Rather, it is because we reach a force balanced between convergence and divergence, an *indeterminate* force. Norton apparently accepts Malament's criticism, since he is moved in his reply (1995) to add a further assumption to (a), (b), and (c), and introduce a somewhat different contradiction, an 'indeterminacy contradiction', as we'll see in §5.3.2.

5.3.1.2 *... using Poisson's equation?* A second method of reasoning involves Poisson's equation. Norton never suggests that we get a contradiction of forces using Poisson's equation directly, but it will be useful later on to consider precisely why the theory isn't inconsistent in this way. Fleshing out assumption (b′) we have:

(b') The force of gravity **F** on a body m_t at **r** due to the mass distribution in a given volume V is given by $\mathbf{F}(\mathbf{r}) = -m_t \nabla \varphi(\mathbf{r})$, where $\varphi(\mathbf{r})$ is such that $\nabla^2 \varphi(\mathbf{r}) = 4\pi G \rho(\mathbf{r})$, where G is a constant and where the gravitational potential $\varphi(\mathbf{r})$ and the mass density $\rho(\mathbf{r})$ are continuous scalar fields on V.

To find the net force on a test mass we no longer need to sum up all the individual forces, but can simply derive the force from the potential field $\varphi(\mathbf{r})$. From assumption (c) ρ is constant in space, so instead of '$\rho(\mathbf{r})$' we can simply write 'ρ', with the added proviso (as we saw in the last section) that the density is only constant for regions of space of a given volume V or greater. Poisson's equation then becomes $\nabla^2 \varphi(\mathbf{r}) = 4\pi G \rho$, where $\varphi(\mathbf{r})$ now refers to the gravitational potential at 'points' of space **r** which actually pick out regions R of volume V or greater. Since Poisson's equation is a differential equation we need to integrate, and when you integrate you inevitably incur constants of integration. Thus the so-called 'canonical solutions' of Poisson's equation are,

$$\phi(\mathbf{r}) = \frac{2}{3}\pi G \rho |\mathbf{r} - \mathbf{r}_0|^2.$$

Here $\mathbf{r}_0$ is the constant of integration.[9] If we differentiate this equation to test whether it satisfies Poisson's equation we get the right result $4\pi G \rho$ whatever $\mathbf{r}_0$ is, because it simply disappears during the calculation.

We can now move to the force on our test mass using $\mathbf{F}(\mathbf{r}) = -m_t \nabla \phi(\mathbf{r})$. We find,

$$\mathbf{F}(\mathbf{r}) = m_t \frac{4}{3}\pi G \rho |\mathbf{r}_0 - \mathbf{r}|.$$

Once again, it should be emphasized that this really means the average force **F** in a large region of space R—picked out by **r**—of volume V or greater. We get a different value for this average force depending on how we choose $\mathbf{r}_0$, so we can get two different, contradictory average forces in the same region R by choosing $\mathbf{r}_0$ in two different ways. But it would be a gross error to go on to suppose that the theory is inconsistent for *this* reason. The theory *just doesn't tell us* what $\mathbf{r}_0$ is, so we must leave it as an unknown constant. We

---

[9] Norton (1995) calls these the 'canonical solutions', and distinguishes them from the most general solutions which include another constant of integration. This other constant is quickly eliminated (see Norton 1995: 513), and at any rate does not affect any argument here (see footnote to §5.3.1.3 for more detail on this point).

certainly can't just arbitrarily choose it to be two *different* things. To emphasize that the theory leaves us guessing we could write $\mathbf{F} = ?$, because $\mathbf{r}_0 = ?$ (this will be useful for comparison later).

5.3.1.3 ... *reasoning from symmetry* Given the tools at our disposal, we've thus far seen two different ways of reasoning when faced with the question, 'What is the net force on a given test particle?' We can use Newton's inverse-square law of gravitation or we can use Poisson's equation. (In both cases a contradiction can be derived only by employing inferences which are *not* truth-preserving.) A third possible method of reasoning, and perhaps the most obvious (particularly to non-scientists), is to use symmetry considerations.

There is an intuition that if the universe is really infinite and Euclidean, and has a homogeneous mass distribution (with the qualifications noted above), then it will be exactly the same vis-à-vis the average force on a test mass in any given region R.[10] In other words, it is assumed that the average force *cannot differ* for any two such regions, since they are identical in the relevant respects. Of course this can't follow from the cosmological assumptions (c) alone, since no reference is made to *force* there. From (c) it only follows that every 'point' of the universe is identical vis-à-vis the surrounding *mass*. So we need to add another assumption to change the intuition into a valid claim that every 'point' in the universe is identical vis-à-vis the gravitational force.

The following assumption will do:

(e) Gravitational force is caused by *all* mass and *only* mass.

If gravitational force is caused by *all* and *only* mass, then the fact that every 'point' in the universe is identical vis-à-vis the surrounding mass distribution will mean that every 'point' in the universe is identical vis-à-vis those factors relevant to the gravitational force. Which will mean that the average gravitational force is identical at every 'point' in the universe. It is often assumed that this means that the force must be everywhere zero. What reason, the argument goes, could there be for the force to point in one direction rather than another? This really needs an additional 'no preferred

---

[10] Focusing on regions rather than points of space obviates the need to account for local variations in the matter distribution, and allows us to compare the results of this method of reasoning directly with that which draws on Poisson's equation.

direction' assumption, which can be ignored for present purposes. All that is required to reach contradiction is that the force, whatever it is, doesn't differ from 'point' to 'point'. Following the discussion in §5.3.1.1, it should be noted that this is also consistent with the force being everywhere indeterminate.[11]

We can now finally achieve a contradiction of forces by comparing this method of reasoning with the one seen in the previous section. From symmetry, drawing on propositions (c) and (e), we have inferred either that $\mathbf{F} = \mathbf{k}$ at every 'point' $\mathbf{r}$ of the universe (for some $\mathbf{k} \in \mathfrak{R}^3$) or that $\mathbf{F}$ is everywhere indeterminate. Either way $\mathbf{F}$ will not differ from 'point' to 'point'. But we saw in the previous section that we can draw on propositions (b') and (c) to conclude that the average force at a given 'point' $\mathbf{r}$ will be

$$\mathbf{F}(\mathbf{r}) = m_t \frac{4}{3}\pi G \rho (\mathbf{r}_0 - \mathbf{r}).$$

In order to satisfy Poisson's equation $\nabla^2 \phi(\mathbf{r}) = 4\pi G \rho$, $\mathbf{r}_0$ *must* be some real vector quantity: $\mathbf{r}_0 \in \mathfrak{R}$. But *whatever* vector quantity we choose we find that the force $\mathbf{F}$ *will* differ from one 'point' $\mathbf{r}$ to another $\mathbf{r}'$. In fact whatever the choice of $\mathbf{r}_0$, we find that the average force on a test mass will be $\mathbf{k}$, for any given $\mathbf{k}$, in exactly one region in the universe. And the difference between regions increases as the distance between the regions increases. Using this fact we reach the following contradiction, resulting from reasoning from (b'), (c), and (e) in two different ways:

(C2) The average force on a test mass in any two arbitrary, widely spaced regions of the universe R (of volume V or greater) will not differ (or differ by a negligible amount), and the average force on a test mass in any two arbitrary, widely spaced regions will differ significantly.[12]

---

[11] Malament (1995: 493 and 509) argues that a homogeneous mass distribution does *not* entail a homogeneous force field. However, this is only when one takes 'gravitational force' to be a gauge quantity with no 'direct physical significance' (as Malament puts it). But before 1900 'gravitational force' certainly was presumed to have physical significance, so the entailment holds true for the purposes of this chapter.

[12] Einstein spent time trying to resolve this particular conflict. He found that the only way to make things work was to alter Poisson's equation to $\nabla^2 \phi - \lambda \phi = 4\pi G \rho$, so that a constant solution for φ is possible: $\varphi = -4\pi G \rho / \lambda$ (he found that no such solution is possible with the normal version of the equation, even if one draws on the most general solutions—see footnote to section §5.3.1.2, this volume). This then gives an average force of F = 0 everywhere, consistent with symmetry considerations. See Norton (1999) for this and other ways in which the assumptions can be modified to avoid the problems.

Although this *is* a contradiction, it is not immediately obvious what it means in empirical terms. We will not find the 'force on a test particle' **F** being, impossibly, both two different things at a single time (in other words we don't get contradiction C1). Whereas the 'force on a test particle' is something we might say 'exists', the *average* force in a given region of space is a non-existent abstraction, just as the average family (with 2.4 children) is an abstraction. The real empirical difference here lies with the large scale movements of matter over time: with one story there are no large scale movements, whilst with the other there will be a large scale acceleration towards the 'point' r_0.

This is closely related to Norton's so-called 'inhomogeneity contradiction', as introduced in his 1995 paper (p.514). However, instead of a contradiction of forces he introduces the contradiction,

(C3) The gravitational potential φ is homogeneous and *it is not the case that* the gravitational potential φ is homogeneous.

This is only achieved by introducing a new assumption:

(d′) The gravitational potential φ is homogeneous.

Unlike the other propositions used so far, commitment to (d′) by scientists is dubious. φ isn't a physical thing after all, but is just a mathematical tool which intermediates between the 'real' masses and forces.[13] The suggestion seems to be that (d′) follows from the homogeneity of the mass distribution, but this is to mix up the physical and the mathematical. The argument is surely that since the universe's mass distribution is symmetrical the universe cannot differ from region to region in certain physical respects. The latter include the (physical) force, but not the (unphysical) potential.

Here it is better to turn to Malament (1995: 492), who frames the difficulty in terms of the homogeneity of the *force* field. Thus a slightly different contradiction is suggested:

(C4) The gravitational field **f** is homogeneous and *it is not the case that* the gravitational field **f** is homogeneous.

---

[13] The potential was introduced by Laplace in the 1770s, and was considered a mere computational tool from the very beginning (see Grattan-Guinness 1990: 332; Cat 2001: 402ff.).

where **f** stands for **f(r)**, the force per unit mass at **r**. This is closely related to (C2), but differs from it in two respects which are worth noting. First, Malament has left out the fact that **f** must stand for the *average* force field in large regions (thus C4 makes things seem simpler than they really are). However, this isn't really a difference since Malament has simply left it implicit. A more substantial difference is that Malament does not allow for the fact that a force everywhere indeterminate is compatible with symmetry constraints (unless, somewhat implausibly, he intends this possibility to be covered by the word 'homogeneous'). In other words one can only reach Malament's contradiction (C4) by drawing on an extra assumption, assumption (d), which blocks the possible indeterminacy of the force. But my analysis shows that drawing on this extra assumption is not necessary to reach contradiction. Leaving assumption (d) aside, from symmetry it follows that *if* such a force field exists *then* it is homogeneous. Poisson's equation then brings contradiction by telling us that such a force field *does* exist, and that it is *in*homogeneous, which affirms the antecedent and denies the consequent of our conditional.

Summing up, what I have shown is that we do get a genuine contradiction of forces (C2) from (b'), (c), and (e). In fact we might even say that we reach the contradiction by (b') and (c) alone, since it might be argued that (e) is embedded within (b'). However, the challenge isn't to find inconsistency in as few propositions as possible. What is more important (as we saw in the Pauli inconsistency) is an inconsistency in propositions which probably were committed to, in the relevant historical period, as serious candidates for the truth. On the face of it, propositions (b'), (c), and (e) meet this criterion. In particular we don't gain anything by eliminating assumption (e) from our assumption set. It was believed to be true by every relevant historical figure in the relevant period: it is part of the real history, and not just an assumption cooked up ad hoc to bring about a contradiction. Second, it is clearly united with the other propositions necessary to reach this contradiction, at least insofar as they all come together to answer question (Q). The significance of this inconsistency will be considered further in §5.4, after we have looked at a different *type* of inconsistency which can be associated with question (Q).

5.3.2 *An indeterminacy contradiction*

The contradiction of concern in this section will be,

(C5) There is a unique gravitational force on a test mass and *it's not the case that* there is a unique gravitational force on a test mass.

This isn't quite the contradiction Norton presents in his 1995 paper, but it is surely what he *means* to present. It is worth pausing to clarify things here, since he doesn't correct his mistake in his 1999 and 2002 papers.

Norton responds to Malament's objections (as seen above in §5.3.1.1) in his 1995 paper. He accepts that there is no contradiction of forces after all, and instead brings to our attention what he calls an 'indeterminacy contradiction'. On p.513 he adds the following assumption to (a), (b), and (c):

(d★) There is a unique gravitational force on a test mass fixed by (b) and (c).

Now, since Malament is right about the non-convergence of the sum, one *cannot* derive a unique gravitational force on a test mass from (b) and (c). Thus it might be supposed that we have a contradiction here:

(C6) There is a unique gravitational force on a test mass fixed by (b) and (c) and *it's not the case that* there is a unique gravitational force on a test mass fixed by (b) and (c).

But on closer inspection we don't have this contradiction after all. If we accept Norton's (d★) we have,

(a) Newton's three laws of motion.
(b) Newton's inverse square law of gravitation.
(c) Matter is distributed homogeneously and isotropically (when viewed on a large enough scale) in an infinite Euclidean space.
(d★) There is a unique gravitational force on a test mass fixed by (b) and (c).

Proposition (d★) gives us the positive contradictory of (C6), so it only matters that we can derive the negative contradictory. However, even if it follows from (b) and (c) that,

It's not the case that there is a unique gravitational force on a test mass.

this *isn't* the contradictory we want. To establish (C6) we need to add '...fixed by (b) and (c)' on the end. But since the propositions in question don't refer to '(b)' and '(c)' at all this is an impossible task. (C6) cannot be derived from (a)–(d★) after all.

It is clear what has happened here. In specifying (d★), Norton has accidentally mixed up the theory and the meta-theory. He actually meant to add,

(d) There is a unique gravitational force on a test mass

which leads to contradiction (C5), as we will see in the next section.

5.3.2.1 ... *using Newton's law of gravitation*   We saw in §5.3.1.1 that, as Malament claims, propositions (a), (b), and (c) tell us that the net force on a given test mass is undetermined. But (d) tells us that the net force on a test mass *is* determined. And the introduction of (d) should not be dismissed as the ad hoc introduction of the required contradictory. In fact, the introduction of (d) is merely the explicit mention of an assumption which is already an integral part of Newton's three laws (a). Take Newton's first law, for example. In its original form it states, 'Every body perseveres in its state of being at rest or of moving uniformly straight forward, except as it is compelled to change its state by force impressed.' This is equivalent to 'a body is either at rest or moving in a straight line, or accelerating due to an impressed force'. These are the only options, so a body is either experiencing a force (accelerating) or it is not (straight line motion or rest). In other words, there is always a determinate force on a body, whether it be something or nothing. Thus the introduction of (d) is not the introduction of a new assumption at all, but is part and parcel of (a). Thus the indeterminacy contradiction (C5) is meant to follow from (a), (b), and (c).

There is an important distinction to make here. Certainly from (b) and (c) we end up with an infinite sum which is indeterminate, from which we cannot achieve an answer to the question 'what is the force?' But can we conclude from here that there is *no* unique gravitational force on a test mass? That is, we have failed using (b) and (c) to determine what the force is. But couldn't it still be the case that there is *some* unique force, and that we could determine what it is by another method, using different reasoning or bringing in other considerations?[14] This is crucial, because if we cannot

---

[14] This is suggested by Norton's introduction of (d★); (d★) says that there is no force *fixed by (b) and (c)*, not that there is no force.

make the further assumption that *no force exists* then we cannot get to the negative contradictory in question, and the inconsistency claim falls down.

The two possible inferences can be distinguished as follows:

(I) from indeterminacy infer that no solution has been reached.
(I★) from indeterminacy infer that there *is* no solution.

If one makes the weaker inference (I) then the reasoning continues as follows. The fact that we can't figure out what the force is from (b) and (c) can be represented by a question mark (cf. §5.3.1.2):

$$\underbrace{\sum_i \mathbf{F}_i = ?}_{(b),(c),(I)}$$

On this understanding *all options are still open*: there might be some other way to determine what the unique force on a test mass is. There is then no contradiction with (d). We might formalize (d) thus:

$$\underbrace{\exists \mathbf{k} \in \mathfrak{R}^3, \mathbf{k} = \mathbf{F}_{net}}_{(d)}$$

In other words, there exists some vector quantity which equals the net force on a test particle.[15] If we believe that (b) Newton's law of gravity holds then we can add,

$$\underbrace{\mathbf{F}_{net} = \sum_i \mathbf{F}_i}_{(b)}$$

Placing these beside each other we have,

$$\underbrace{\exists \mathbf{k} \in \mathfrak{R}^3, \mathbf{k} = \mathbf{F}_{net}}_{(d)} \quad \underbrace{\mathbf{F}_{net} = \sum_i \mathbf{F}_i}_{(b)} \quad \underbrace{\sum_i \mathbf{F}_i = ?}_{(b),(c)(I)}$$

Now, by substitutivity of identicals, we can fill in the question mark and write:

---

[15] Of course, before the introduction of 'real numbers' and the like, physicists would have talked vaguely about 'quantities', but that doesn't affect the argument at hand.

$$\underbrace{\exists \mathbf{k} \in \mathfrak{R}^3, \mathbf{k} = \sum_i \mathbf{F}_i}_{(b),(c),(d)}$$

There is no contradiction here. We couldn't find an answer to our indeterminate sum using (b) and (c), but there is *some* answer, yet to be discovered.

Today, two centuries of mathematics tells us that this is the wrong way to think about indeterminate sums. Not only do we *get* no answer when we are faced with an indeterminate sum, we find that there *is* no answer, there *cannot* be an answer, as stated by (I★).[16] As far as the sum of gravitational forces goes, this means that,

$$\underbrace{\forall \mathbf{l} \in \mathfrak{R}^3, \sum_i \mathbf{F}_i \neq \mathbf{l}}_{(I\star)}$$

With this in place we really do have our contradiction. We now have the following three equalities:

$$\underbrace{\exists \mathbf{k} \in \mathfrak{R}^3, \mathbf{k} = \mathbf{F}_{net}}_{(d)} \quad \underbrace{\mathbf{F}_{net} = \sum_i \mathbf{F}_i}_{(b)} \quad \underbrace{\forall \mathbf{l} \in \mathfrak{R}^3, \sum_i \mathbf{F}_i \neq \mathbf{l}}_{(b),(c),(I\star)}$$

From the substitutivity of identicals we can then write,

$$\exists \mathbf{k} \in \mathfrak{R}^3, \forall \mathbf{l} \in \mathfrak{R}^3, \mathbf{k} \neq \mathbf{l}$$

To be logically rigorous, we could now perform existential and universal instantiation to reach the conclusion $\mathbf{a} \neq \mathbf{a}$. In other words it follows that some three-vector $\mathbf{a}$ is not equal to itself, a blatant contradiction.[17]

Thus the inference we make when faced with an indeterminate sum decides whether we derive a contradiction or not. As we will see further in §5.4.2, the force of inference (I★) can be easy to overlook, and has been overlooked by several authors both in the distant and recent past. In fact

---

[16] Cauchy wrote in 1821, '*a divergent series does not have a sum*' (p.iii). However, in this chapter I am assuming only that *alternating* divergent series do not have a sum. The reason is that, if we are talking *physics* (rather than mathematics) we cannot infer that a quantity which 'diverges to infinity' is equal to *nothing*. Infinity counts as 'something' here.

[17] We wouldn't avoid inconsistency if we introduced a new type of number (as is sometimes done in mathematics), the 'indeterminates', so that the alternating divergent series in question *does* have a sum. It still wouldn't have a physically relevant sum (e.g. it wouldn't equal any *real* number), and this is enough to conflict with Newton's laws.

what (I*) tells us is that absolute forces in the universe *do not exist* (that our metaphysics is wrong).[18] The point is that one must have assumed that absolute forces exist in order to ask the question 'What is the net force on a test particle?' But then when we employ (I*) we find that absolute forces do *not* exist, contradicting the assumed metaphysics. So when one reaches indeterminacy in the way seen here, it is really just another way of reaching contradiction. This puts some meat on the bones of Malament's suggested distinction (1995: 489) between the theory being inconsistent and the theory being 'unacceptable a priori' (because of indeterminacy): the latter is a special case of the former.

5.3.2.2 *...from summing the potential* φ There is one final method of reasoning we have not yet considered, and there is something of a tradition of using it to demonstrate the failures of 'the theory'. The gravitational potential at a point **r** due to a given mass m_i at $\mathbf{r}_i$ can be expressed thus:

$$\phi_i(\mathbf{r}) = -G \frac{m_i}{|\mathbf{r}_i - \mathbf{r}|} \tag{5.1}$$

The net gravitational potential at a point will then be $\phi_{net}(\mathbf{r}) = \sum_i \phi_i(\mathbf{r})$. This then gives rise to a new assumption about forces in the universe:

(f) The net force of gravity $\mathbf{F}_{net}$ on a body m_t at **r** is given by $\mathbf{F}_{net}(\mathbf{r}) = -m_t \nabla \phi_{net}(\mathbf{r})$, where ϕ_{net} is achieved by summing the gravitational potential (equation (5.1)).

But the problem with this method of reasoning is that the net gravitational potential is everywhere infinite: whereas the components of *force* on two opposite sides of a test mass, being vector quantities, cancel each other out (to one degree or another), the components of potential, being scalar quantities, accumulate. This time not only does the sum diverge to infinity, but it diverges to infinity relatively quickly because the potential is a $1/r$ relationship, whereas masses in the universe increase with r^2.[19]

---

[18] This is the conclusion of Norton's 1995 paper (see especially p.515). Note the subtle difference between saying that there are no forces (i.e. F = 0 everywhere) and saying that absolute forces don't exist. In the latter case 'F' doesn't refer, so it can't equal anything, including zero.

[19] Cf. Norton (1999: 273). Grandi's series arises in the context of Newtonian cosmology for the *force* because the force is a $1/r^2$ relationship and masses in the universe increase with r^2, thus cancelling each other out.

What should we conclude from the fact that the potential diverges to infinity at every point? As noted in §5.3.1.3, the potential is merely a mathematical tool, used to mediate between *physical* masses and forces. If there is trouble in an infinite potential, that should only be in the fact that the *physical* consequences are unpalatable. Now, as stated, to reach the force on a test mass m_t from the potential we need to take its gradient: $\mathbf{F}(\mathbf{r}) = -m_t \nabla \phi(\mathbf{r})$. But if φ is infinite everywhere this operation isn't possible, because it is not defined for infinity. One cannot proceed to derive $\mathbf{F}(\mathbf{r}) = \mathbf{0}$, reasoning that $\varphi(\mathbf{r})$ is everywhere *constant*. 'Constant' refers to numerical constancy, and infinity is not a number. Thus, I suggest, not only do we find that we don't know what $\mathbf{F}$ is in this case, we find that $\mathbf{F}$ is *indeterminate*. Thus summing the potential is consistent with summing the force directly using Newton's law of gravity (b), as in §5.3.2.1. So this is really just the same problem of indeterminacy in a different guise.

There have been several related discussions, but nowhere has the problem been identified with indeterminacy. Jaki (1969) writes of a 'gravitational version of Olbers' paradox' where in the latter case, in an infinite, homogeneous universe, the light from distant stars accumulates to give an infinite amount of light at any point (see §8.3 of Chapter 8, this volume).[20] In the gravitational case Jaki (1979) writes,

An infinite universe of homogeneously distributed stars or galaxies cannot exist because in such a universe the gravitational potential is infinite at any point. (Jaki 1979: 121)

But unlike light the potential is a non-physical thing. Nowhere do we find a discussion of exactly *why* an infinite potential is impossible; nowhere is there a discussion of indeterminacy. And those whom Jaki draws on apparently think that the theory demands genuinely infinite *forces*. In particular, Jaki draws at length on Einstein (1917), who argues as follows:

According to the theory of Newton, the number of 'lines of force' which come from infinity and terminate in a mass m is proportional to the mass m. If, on the average, the mass-density ρ_0 is constant throughout the universe, then a sphere of volume V will enclose the average mass $\rho_0 V$. Thus the number of lines of force passing through the surface F of the sphere into its interior is proportional to $\rho_0 V$.

---

[20] Olbers, in 1823, derived that the light at any point in the universe would be equal to k(1+1+1+1+...) for a given constant k (see Jaki 1969: 134f.). Clearly this is not indeterminate: it diverges to infinity.

For unit area of the surface of the sphere the number of lines of force which enters the sphere is thus proportional to $\rho_0 \frac{V}{F}$ or to $\rho_0 R$. Hence the intensity of the field at the surface would ultimately become infinite with increasing radius R of the sphere, which is impossible. (Einstein 1917: 106)

This is worth quoting in full, because to my knowledge it has not yet been made clear that this reasoning is seriously incomplete. How are the final words 'which is impossible' warranted? Rather than focusing on the potential φ Einstein is here focused on the force field **f**, the force per unit volume at a point. It is certainly true that, in our infinite universe, there will be an infinite *component* of **f** in a given direction, but what is so impossible about this? The impossibility only comes when we consider the combined effects of all such infinite components and find that the result is indeterminate, as in §5.3.2.1, above. The impossibility does not lie simply in the absurdity of an 'infinite force field', as Einstein suggests.

The supposition that it is possible to derive infinite forces is not unique to Einstein. Seeliger, who finally shed light on the problems with Newtonian cosmology in 1895, supposes that there are infinite forces in a follow-up paper of 1896. He writes, 'It follows from potential theory that there must be in the *universe unlimited (infinitely) great accelerations*' (cited in Norton 1999: 279, emphasis in original). But this simply isn't the case. Once again, all that is shown is that there will be an infinite *component* of force in a given direction, or that the *potential* will sum to infinity. Kelvin is similarly unclear in 1901 (see Norton 1999: 285).

In summary, we *do* get a contradiction here from summing the potential, and it is once again the indeterminacy contradiction (C5). This time we reach the conclusion because the gradient of a scalar field which is everywhere infinite is indeterminate just as the sum of an alternating divergent series is indeterminate. So what we have here is not a 'qualitatively different' type of problem, as Norton (1999: 279) claims, but just a different way of reaching the same conclusion.

5.3.2.3 ... *using Poisson's equation*   Norton claims that the indeterminacy contradiction (C5) also follows from applying Poisson's equation:

The addition of the potential φ and Poisson equation does not materially affect the indeterminacy contradiction of Newtonian cosmology. There are as many canonical solutions as there are choices for $\mathbf{r}_0$. Each distinct choice of $\mathbf{r}_0$ leads to a different force on the test body. (Norton 1995: 514)

So Norton is claiming that indeterminacy follows from the fact that, depending on how we pick the constant of integration, we get a different result for the force. So no unique force follows from the theory, just as no unique force followed when we had an infinite sum in §5.3.2.1. But here as before we need to make a distinction between no unique force *following from the theory* and there being no unique force *at all*. In §5.3.2.1 this was expressed as two 'strengths of inference' (I) and (I★). With Poisson's formulation we get an analogous pair of inferences:

(II) From an unknown constant of integration infer that the theory provides no unique solution.
(II★) From an unknown constant of integration infer that there *is* no unique solution.

But this time only the weaker inference (II) is legitimate. This is because the reason why one cannot infer two contradictory forces is different. Recall that in §5.3.1.1 we couldn't infer two contradictory forces because there *cannot be* a solution to an indeterminate sum. But in §5.3.1.2 we couldn't infer contradictory forces because, although there certainly *can* be a solution to an equation with an unknown constant, the relevant propositions couldn't tell us what it was. Since we cannot make the stronger inference we cannot reach the contradictory in question and, *contra* Norton, there is no indeterminacy contradiction here.

However, when we compare the method of reasoning based on Poisson's equation (b′) and that based on Newton's law of gravitation (b) we *do* find a conflict. We saw above that if one applies (b) to an infinite homogeneous universe, one infers that the force is indeterminate. But, as seen in §5.3.1.2, using Poisson's equation instead we can move from $\nabla^2 \phi(\mathbf{r}) = 4\pi G \rho$ to

$$\phi(\mathbf{r}) = \frac{2}{3}\pi G \rho |\mathbf{r} - \mathbf{r}_0|^2,$$

where $\mathbf{r}_0$ must be some real number (it cannot be indeterminate since then it would not be a solution to Poisson's equation). And from here, using $\mathbf{F}(\mathbf{r}) = -m_t \nabla \phi(\mathbf{r})$, since one has a determinate potential one has a determinate force. So from (b), (b′), and (c) we can infer the indeterminacy contradiction (C5) once again, where this time the determinacy of the force follows from (b′).

This seems to go against Malament (1995). He *seems* to claim that (b′) is actually a generalization of (b) because when the infinite sum in question converges (b) and (b′) agree, whereas (b′) can also be applied when the sum doesn't converge. He writes,

> There is a clear sense in which it [the 'differential' formulation] is a generalization, with a wider domain of application.... The 'integral' formulation is not applicable to cosmological contexts of the sort we have considered. (Malament (1995: 491 and 508)

But if (b′) were a generalization of (b) then they would not be in conflict. And they are in conflict, as I have argued above.

However, it turns out that what Malament calls the 'integral formulation' and the 'differential formulation' are not quite the same as (b) and (b′). They are the same except for one crucial interpretational difference: for Malament *gravitational force is taken to be a gauge quantity without direct physical significance*. But (b) and (b′) were around long before this twentieth-century attitude to 'gravitational force'. Before 1900 (b′) was *not* a generalization of (b), and they were in fact in conflict. Thus Malament's analysis and the above analysis can stand side by side: one need only note that Malament is concerned with twentieth-century developments of Newtonian cosmology, and I am not.[21]

5.4 Why weren't the Inconsistencies Noticed?

In all this we find four notable inconsistencies:

(I1) C2 follows from (b′), (c), and (e). (§5.3.1.3)
(I2) C5 follows from (a), (b), and (c). (§5.3.2.1)

---

[21] I doubt Malament would also claim that (b), like the 'integral formulation', 'is not applicable to cosmological contexts of the sort we have considered' (that is, when the sum does not converge). As has been shown above, one certainly can apply (b) in such a context, with the result that the force on a test mass is indeterminate. To maintain that this means that (b) is inapplicable would be to use one's assumptions like a 'toolbox', where one picks and chooses one's assumptions depending on whether they lead to desirable results. Certainly *some* scientific practice proceeds in this way, but only when the assumptions one trades in are not being considered as candidates for truth.

(I3) C5 follows from (c), (d), and (f). (§5.3.2.2)
(I4) C5 follows from (b), (b′), and (c). (§5.3.2.3)

Thus it might seem perfectly natural to declare 'Newtonian cosmology was riddled with inconsistency!' One might then go further and declare this 'yet another example of inconsistency in science', and go on to consider how scientists reason in the face of inconsistency, perhaps in an attempt to motivate one or another paraconsistent logic, or an account of 'inferential restrictions' in science.[22]

But in fact the historical and scientific details of Newtonian cosmology show us (once again) the non-uniformity of science. This inconsistency is unlike any of those seen so far, in substantial ways, and such that any substantive general claims about 'inconsistency in science' are bound to fall flat. Theory eliminativism brings this non-uniformity to the fore: although 'Newtonian cosmology' isn't usually considered to be a 'theory' in the traditional sense, we are nevertheless in the habit of using the noun-term 'Newtonian cosmology' as if there is this one, single thing with unambiguous content that we are talking about, for which the answer to the question 'Is it inconsistent?' will be 'Yes' or 'No'. The truth is much more complicated, and is best revealed by talking in terms of inconsistent sets of propositions (as summarized above), and the historical relevance and scientific character of those propositions. In this way it becomes clear why the inconsistencies were not noticed until 1895. I start in the next section with historical relevance, before considering an important feature of the 'scientific character' of the propositions in the following section.

5.4.1 Because the right question wasn't asked

The propositions relevant to the inconsistencies are clearly relevant to the real history of science to *some* degree or another. Each of the propositions enjoyed serious commitment, for *certain periods* at least, between the years 1700 and 1900, and many were widely regarded as obvious truths. For example consider assumption (c): Bertrand Russell was convinced enough to write in 1897 that the infinitude and homogeneity of the universe, far from being working hypotheses (say), were scientific principles 'established

---

[22] Cf. Davey (2003).

forever', as Jaki (1969: 184 and 220) puts it. In addition Jaki (1969) writes as follows:

Tammann wrote that since Newton's time 'many astronomers had assumed this idea [the infinitude of the universe] as a working hypothesis.' As a matter of fact, they assumed it as an unquestionable truth. (Jaki 1969: 85)[23]

Then consider propositions (a) and (b), Newton's three laws of motion and his law of gravitation. Clearly these were widely regarded as true, or at least as very serious candidates for the truth, by the relevant people in the relevant time-period. And the other relevant propositions—(b′), (d), (e), and (f)—enjoyed a similar commitment for at least *some* of the relevant time period. There is certainly the possibility, then, of a serious and important inconsistency comparable to the Pauli inconsistency we looked at in Chapter 3.

On the other hand, the caveat 'for *some* of the relevant time period' is important, and makes these inconsistencies very different from the Pauli inconsistency. Consider again assumption (c): matter in the universe is distributed homogenously in an infinite Euclidean space. This assumption features in all four of the inconsistencies, but suppose we ask the question 'Which scientists were committed to it?' For the nineteenth century (up until 1895 at least), there is some sense in saying that nobody was committed to it, either doxastically or for the purposes of calculation. The fact is that, during this period, the relevant individuals didn't ask cosmological questions *at all*. Merleau-Ponty (1977: 283) refers to 'the disappearance of cosmological science as such in the nineteenth century, that is, the investigation of the properties of the Universe considered in its totality—until its surprising revival in the twentieth century.' It is this 'revival' which explains the title of his book of 1976 (co-written with Morando): *The Rebirth of Cosmology*. Therein he goes as far as to say that, in the nineteenth century, 'cosmology itself no longer existed' (Merleau-Ponty and Morando 1976: 66).

This is a remarkable claim, since there was certainly much work in astronomy and celestial mechanics during this period. But, regarding the former, 'in the course of the [nineteenth] century astronomers were discussing the nature and internal structure of individual nebulae rather than

---

[23] For primary evidence of Jaki's claim here—and for further historical details generally—I refer the reader to his excellent book (Jaki 1969).

the wider cosmological problem' (North 1965: 16. Cf. Merleau-Ponty 1977: 291 and Jaki 1979: 117). Similarly those working in celestial mechanics, such as Poisson himself, avoided cosmology entirely. For example Laplace, one of the founding fathers of celestial mechanics and active in the late eighteenth and early nineteenth centuries, never made even a single conjecture as to the structure of the universe as a whole (Jaki 1969: 98; Merleau-Ponty 1977: 283). And later in the nineteenth century, as Jaki puts it,

> The silence of Urbain J. J. Leverrier, the most celebrated French astronomer of those times ... illustrated the typical aversion to cosmological problems on the part of most skilful experts on celestial mechanics. (Jaki 1969: 157)

When this is brought to light, the absurdity of the claim 'In the nineteenth century, Newtonian cosmology was inconsistent' is revealed. No scientist had the relevant inconsistent belief set in that period (at least in his or her capacity *as a scientist*), and obviously nobody was reasoning with the inconsistent propositions. Or, if they were, it was a single individual in a private moment for the purposes of personal curiosity, from which we obviously cannot draw any interesting lessons about 'how science works'. But this is just the nineteenth century; what of the eighteenth century? Merleau-Ponty and Morando's (1976) book is called *The Rebirth of Cosmology*; when did 'cosmology' exist before?

Certainly before the nineteenth century cosmological questions were more popular. Let's suppose for now that assumption (c) was committed to in the eighteenth century by a significant number of relevant individuals. But then we need to consider some of the other propositions necessary to reach inconsistency. For example, commitment to assumption (b′) is absolutely non-existent in the eighteenth century, because Poisson's equation was only introduced in 1813. Similarly, no serious commitment to assumption (f) will be found in the eighteenth century by those interested in cosmological questions. Recall that Laplace introduced the potential in the 1770s, and 'potential theory' gradually developed from there.

So at this stage we see that inconsistencies (I1), (I3), and (I4) do not do any justice to how the history of science really played out. For each one, and for each of the two centuries in question, there is at least one assumption that didn't have any relevance to the real history of science, either because the relevant mathematics didn't exist (roughly speaking before 1800) or because cosmological questions weren't being asked at all (after 1800).

This leaves inconsistency (I2) in the eighteenth century. In this period Newton's laws of motion and his law of universal gravitation were universally accepted, and cosmological questions were being asked by some. Textual evidence exists that individuals committed to assumption (c).[24] So for *this* period we may ask why the inconsistency of propositions (a), (b), and (c) wasn't noticed, or wasn't acted upon. At this point it is helpful to talk in terms of questions being asked. I've said that nobody was asking questions about cosmology in the nineteenth century. But in the eighteenth century there is good reason to believe that nobody, or hardly anybody, ever asked the *right* question. The above analysis highlights one reason for this: the relevant question is actually rather obscure. This is made obvious by the contradiction of forces (C2) of §5.3.1.3. We are not asking what the *actual* force on a given body is: to answer this one would need to know, absurdly, the positions and masses of an infinite number of bodies. Rather, our question (Q) needs to be changed to,

(Q′) What is the *average* net gravitational force a test body would experience over all points of an arbitrary region of the universe R of a given volume V large enough so that the universe is homogeneous at that scale?

This could also be framed in terms of the force field **f**, as per Malament (§5.3.1.3), but still we would not get away from the complications of averaging. And, complications aside, it isn't immediately clear why this question is an interesting one, *except* that answering it in two different ways leads to inconsistency.

Thus even if propositions (a), (b), and (c) were all committed to by relevant individuals in the eighteenth century, one might doubt whether they were all committed to *together*, consciously, at the same time, as part of the same enquiry. Question (Q) acts to bind them together, to make them 'pointedly grouped propositions' in the sense that they all come together to answer that question. We can *now* ask that question, but if it was never asked at the time then the propositions would never have 'come together' scientifically. Thus to say that there is an inconsistency *in science* here (as Norton, Davey, and others have) is liable to be misleading. On the other hand there *is* a set of inconsistent propositions all of which were committed to by

---

[24] Jaki (1969) is a good source of references for this textual evidence.

relevant individuals as serious candidates for the truth (even if the propositions hardly ever 'came together' as part of a scientific enquiry, because nobody ever asked a question which required this). So we may still wonder why the inconsistency apparently wasn't noticed in the eighteenth century, especially by Newton.

5.4.2 *Because of confusion about non-convergent series*

We are now only considering the eighteenth century, and before the nineteenth century is before Cauchy. Since inconsistency (I2) leads to the indeterminacy contradiction (C5), appreciating it depends on making the right inference when faced with a non-convergent series. Thus there is some reason to suppose that a lack of understanding of the relevant mathematics contributed to the inconsistency going unnoticed.[25] In general terms we may say that one of the inferences necessary for the derivation of the contradiction was a peculiar type of inference, alien to the relevant individuals. More specifically, I will provide some evidence in this section that certain individuals made inference (I) rather than inference (I★), as introduced in §5.3.2.1 and repeated here for convenience:

(I) from indeterminacy infer that no solution has been reached.
(I★) from indeterminacy infer that there *is* no solution.

This will also constitute my preferred explanation of the attitudes of those—Isaac Newton and Svante Arrhenius—who, according to Norton, favour a 'no-solution needed' solution to the inconsistency. He characterizes this attitude as follows:

They are aware of the inconsistency but ignore the possibility of deriving results that contradict those that seem appropriate.... At first glance, it would seem that the physical theorists avoid logical anarchy by the simple expedient of ignoring it! (Norton 2002: 191)

Instead I claim that they *weren't* aware of the inconsistency after all, because they only made inference (I)—'no solution reached'—and not (I★)—'no

---

[25] Of course (I3) and (I4) also lead to the indeterminacy contradiction, so this might also explain why they were not appreciated, at least in the early decades of the nineteenth century (before Cauchy but after Poisson's equation and potential theory were established).

solution possible'. To decide between Norton's claims and my own a look at the primary evidence is required.

First to Newton. Was he aware of the inconsistency and chose just to ignore it, as Norton claims? In fact, although Norton does describe Newton as subscribing to a 'no-solution solution' in his 2002 paper, in his historically focused 1999 paper he suggests instead that Newton *wasn't* aware of the inconsistency. When the theologian Richard Bentley pressed Newton on the gravitational consequences of an infinite universe in 1692, Newton referred to how mathematicians handle infinities in terms of limits and convergence. Thus Norton concludes,

> Having recalled for us that there are perfectly good methods of comparing infinites by means of limits, Newton seemed not to have applied them himself to the problem at hand.... It is hard to understand how Newton could make such a mistake. His mathematical and geometric powers are legendary. Perhaps Newton was so sure of his incorrect result from the symmetry considerations that he did not deem it worthwhile the few moments reflection needed to see through to a final result. (Norton 1999: 290f.)

This story goes against the 'no-solution solution' as described in his 2002 paper. He further writes that Newton 'would surely have noticed' that there was an inconsistency if only he had applied the relevant mathematics. So it is *not* the case that Newton noticed the inconsistency but chose to ignore it, as per the 'no-solution solution'.

This suggests that we read Norton in another way. In sum he appears to be saying that *either* (i) Newton didn't apply the relevant mathematics and so didn't notice the indeterminacy, or (ii)—the 'no-solution solution'—Newton *did* apply the relevant mathematics, noticed the indeterminacy, and chose to ignore it.

The present analysis can offer an alternative explanation. The fact that Newton was clear on how to handle *converging* infinite series is actually irrelevant insofar as Newtonian cosmology is concerned. The relevant series is infinite and *diverging*, so Newton *couldn't* have applied the methods of limits to the problem at hand (as Norton suggests). And in fact Newton's grasp on divergent series, and alternating divergent series in particular, was not good. In his most in-depth writings on infinite series[26]—an unpublished

---

[26] See Whiteside's annotation in Newton (1981: 267).

essay from 1684 entitled 'On the computation of series'—Newton blatantly overlooks the fact that a divergent alternating series has no sum. Following one particular passage Whiteside's annotation reads,

> He has, however, ignored the unpleasant fact that no unique sum is assignable to a divergent alternating series. (Newton 1971: 611)

Whiteside is using the word 'ignore' in a loose sense here, and doesn't mean to suggest that Newton saw the correct conclusion but decided to ignore it. Newton was not in the habit of ignoring what he knew to be correct conclusions.

In sum, then, a 'third way' seems a more plausible explanation of Newton's attitude than either of Norton's suggestions. This is to suppose that, if Newton *did* make the calculation in question, upon coming across an alternating, divergent series he made inference (I)—no solution reached—rather than inference (I★)—no solution possible. If he *found* no solution, but didn't conclude that there *was* no solution, he would have tried a different tack. As Norton notes, 'symmetry considerations' may have guided him, so that he concluded that the average net force must be zero (cf. §5.3.1.3, above).[27] The infinities must balance after all, although apparently mathematics isn't up to the task of showing us this.[28]

To give a second example, Norton writes that Arrhenius 'laid out a clear statement of the "no-solution solution".' Arrhenius wrote in 1909,

> [I]t is very much understandable that Seeliger's argumentation is frequently construed as conflicting with the infinity of the world. *This, however, is not true.* The difficulty lies in that the attraction of a body surrounded by infinitely many bodies is undetermined according to Seeliger's way of calculation and can take on all possible values. This, however, only proves that one *cannot* carry out the calculation by this method. (cited in Norton 1999: 291, emphases added)

---

[27] Jaki (1969: 60–5) gives a nice discussion.

[28] Another possibility is that Newton would have said that the indeterminacy is an artefact of the mathematics because unidirectional vector representations of force are mathematically convenient representations of physical interactions between bodies which are in fact inseparably mutual. (I owe this thought to Hylarie Kochiras; cf. Kochiras 2013.) In other words, the maths doesn't exactly represent what is happening physically. This would mean that, for Newton, assumptions (a) and (b) are *not* candidates for the truth after all, because of the unidirectional vector variables in them. In fact I doubt that the inconsistency can really be avoided in this way, but in any case most of Newton's followers would not have followed Newton's philosophy here. Nevertheless this 'artefact of the mathematics' justification for ignoring inconsistency will come up again, and is worth keeping in mind. See §8.2.4, this volume.

Certainly Arrhenius did *not* think that there was an inconsistency, as Norton suggests. The confusion here seems to rest with what Arrhenius means by 'cannot'. As Norton interprets it, when Arrhenius says that 'one cannot carry out the calculation by this method' he means that, although mathematically sound, one must *avoid* that method of calculation because it leads to contradiction. This, however, leaves inexplicable why Arrhenius thinks there is no conflict. Things make more sense if we read 'one cannot carry out the calculation by this method' in another way. Arrhenius means that one just doesn't get an answer that way—one does not *reach a solution* by this method—because the sum in question is indeterminate. But this means that there still may *be* an answer, and he suggests zero (based, again, on symmetry considerations). His mistake is in not making as strong an inference as he ought to make when faced with non-convergence (he makes inference (I) instead of (I★)).

Even as late as 1954 Layzer, criticizing Milne and McCrea's neo-Newtonian cosmology of the 1930s, made this same oversight and claimed that we should infer that **F** = **0** everywhere (Layzer 1954: 269). McCrea put things straight the following year:

[I]f the gravitational force is to be defined in the present manner, then *it does not exist* in the case of uniform density. Accordingly, nothing further can be inferred about this case. In particular, we may not proceed to argue, as Layzer does, that the force must be the same at every point, and thence that it must be zero. For, in order to prove that a force takes *any* value, in particular the value zero, the force has to exist in the mathematical sense. (McCrea 1955: 273, emphasis added)

What he surely means by the final remark is that, if the force really is equal to an indeterminate sum, then it can take *no* value, including zero.

Still other authors, who clearly do understand non-convergence very well, aren't sufficiently careful with their words to make the distinction between the force being zero (it is determined) and the force not existing (it isn't determined). Even parts of Malament (1995) are unclear on this point. In his criticism of Norton (1993) he writes,

The integral I is not convergent, and so it is *not* the case that $I = I_1 + I_2 + I_3 + \ldots$ [...]. Newtonian theory...makes no determination [of gravitational force] at all. (Malament 1995: 491, original emphasis)

Here we have two clear statements of the *weaker* of our two inferences (I). The stronger inference (I★) would state not only that Malament's integral 'I' is not equal to $I_1 + I_2 + I_3 + \ldots$ but that it is equal to *no quantity whatsoever*. And it would state not only that Newtonian theory *makes no determination* of the net gravitational force, but that Newtonian theory states that the net gravitational force cannot *be* any quantity. Compare this with §5.3.1.2, where Poisson's equation makes no determination of the net gravitational force (because we are left with unknown constants of integration), but nevertheless demands that it is *some* quantity.[29]

This subtle distinction between no force being *found* and no force being *possible* (even zero) is just the tip of the iceberg when it comes to confusion about divergent series in the relevant period, especially in the eighteenth century. Euler and others set the sums of divergent series equal to certain quantities right through the eighteenth century, and were able to reach startling *correct* conclusions by manipulating them (see Hardy 1949: ch. 1). That is, setting divergent summations equal to certain values proved to be extremely fruitful. In addition, as Hardy explains,

> [T]here is only one sum which it is 'reasonable' to assign to a divergent series: thus all 'natural' calculation with the series [1−1+1−1+ . . .] seems to point to the conclusion that its sum should be taken to be ½. (Hardy 1949: 6)

Apart from the fact that ½ is the mean of 1 and 0, there were some very persuasive reasons to set Grandi's series equal to ½. For example if we set

$S = 1 - 1 + 1 - 1 + 1 - 1 + \ldots$

then we might conclude that,

$1 - S = 1 - (1 - 1 + 1 - 1 + 1 - \ldots) = 1 - 1 + 1 - 1 + 1 - \ldots = S.$

This would mean that $1 = 2S$, or that $S = ½$. Another method was to consider the binomial expansion (discovered by Newton in the 1660s),

$$\frac{1}{1+x} = 1 - x + x^2 - x^3 + \ldots$$

---

[29] Further, Malament's statement that Newton's law of gravitation is 'not applicable' when the series in question doesn't converge (see §5.3.2.3, above) also suggests the weaker inference (I). And recall also Norton's introduction of assumption (d★), rather than (d), which strongly suggests the same.

This was known to converge for all x such that $0 \leq x < 1$. But if it holds for all such x then in the limit at x goes to 1 we find that,

$$\frac{1}{2} = 1 - 1 + 1 - 1 + \ldots$$

This latter method had already been recommended by Leibniz and was still popular one hundred years later, in the early nineteenth century. Poisson himself favoured this reasoning, despite living in the time of Cauchy's groundbreaking *Cours d'Analyse* of 1821, and he 'retained this staple component of his analysis throughout his life' (Grattan-Guinness 1970: 88; see also Laugwitz 1989: 218ff.). In fact, Grattan-Guinness claims that when Cauchy wrote in italics '*a divergent series has no sum*' (1821: iii) it was partly aimed at Poisson (Grattan-Guinness 1970: 88).[30] And even as late as 1844 De Morgan still failed to appreciate that Grandi's series did not sum to ½ (see Hardy 1949: 19f.).

So in conclusion we see that the inconsistency (I2)—and indeed any inconsistency leading to the indeterminacy contradiction (C5)—depends on appreciating a distinction which is very subtle in language, but not at all subtle in meaning. The problem is that saying 'there is no force' can very reasonably be taken to mean that the magnitude of the force is zero, rather than that there is no such thing as a 'force'. But if we find that the 'force' is indeterminate, and we trust our mathematics, then we must infer that there is simply no such thing as a net force on the given body. Until some time after Cauchy entered the scene in 1821, it was difficult to appreciate that this inference was truth-preserving. And thus it was very easy to fail to notice the inconsistency in one's assumptions.

5.5 Conclusion

The main conclusions of this chapter can be summed up as in Figure 5.2. We find that some of the relevant propositions are not historically relevant because they just didn't exist in any sense at that time: here I am thinking of Poisson's equation and potential theory in the eighteenth century. Other propositions are not historically relevant because they simply weren't being

---

[30] For more on Poisson and divergent series, see Grattan-Guinness (1990: 731f).

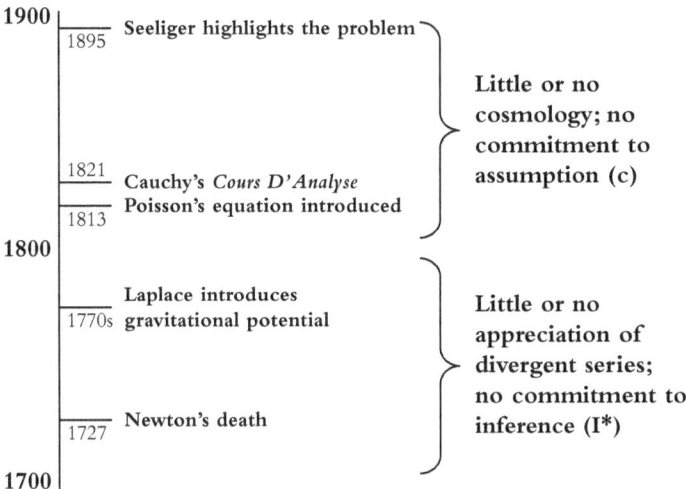

Figure 5.2. A timeline showing some key factors relevant to the inconsistencies associated with 'Newtonian cosmology'

considered by the relevant individuals at that time: here I am thinking of the cosmological assumptions (c) in the nineteenth century. Inconsistency (I2) avoids these problems of historical relevance to a large extent because it doesn't draw on Poisson's equation or potential theory. Thus it has some genuine relevance in the eighteenth century at least. Here I have claimed that the inconsistency wasn't noticed because one particular truth-preserving inference (concerning divergent series) necessary to derive the contradiction was very hard to appreciate until after the work of Cauchy and others in the 1820s.

There are several things we can infer from this particular example. First of all there is the obvious point that there may be important lessons to be learnt from applying one's scientific hypotheses in a new context. The lull in cosmological questions in the nineteenth century in particular meant that the inconsistency in question was highly likely to go unnoticed. But more important than this is the point that the *particular* cosmological question which had to be asked to notice the inconsistency is not an especially obvious question to ask. When it takes the form of (Q) it just seems silly to ask it, because it obviously can't be answered. But when it takes the form of (Q′) it doesn't seem like an *interesting* question. What needs to be appreciated here is that questions which do not appear to be particularly interesting may have more interesting answers than might be supposed. In this case the question (Q′) could be answered in two different, contradictory ways, showing scientists one of the most important things they can learn: that one of their assumptions is definitely false. This is all the more striking when the propositions are as solidly entrenched as these were.

Recall that I came to a similar conclusion in the case of the later Bohr theory: it equally wasn't clear why Pauli's question about how magnetic and electric fields would affect the hydrogen atom was interesting. But of course it became *very* interesting when he derived a contradiction. Another comparison between this case and that of the later Bohr theory is worth making here: in both cases the inconsistency is somehow hidden in the mathematics. However, the reason why it is hidden is not the same. The main reason in the Bohr case is that the derivation of the contradiction was long and complicated, as a glance at Pauli's original paper will confirm (or see Mehra and Rechenberg 1982: 507–9). But in Newtonian cosmology the derivation is not nearly so complicated. This time the difficulty lies entirely in appreciating that a certain inference—inference (I★)—is truth-preserving.

So in the end how important is the inconsistency in 'Newtonian cosmology'? I certainly don't see value in a scale of importance: I've already noted some of the ways in which the inconsistency of propositions (a)–(c) is qualitatively different from the Pauli inconsistency, for example. On the other hand, both inconsistencies are very serious in the sense that scientists ultimately had to change their views about propositions which had previously been considered candidates for the truth. In the Pauli case I said that this wasn't so hard: the '$m \neq 0$' assumption could be dropped relatively easily. In the Newtonian cosmology case things were much more difficult.[31] When Seeliger introduced the inconsistency in 1895 very many suggestions were made as to how consistency might be restored.[32] With the benefit of hindsight (with general relativity now in hand) one can suggest that the final, best fix was to move from absolute acceleration to relative acceleration (see Norton 1995). In a sense, this is to adopt an anti-realist 'gauge' interpretation for accelerations and forces (cf. Malament 1995), in a manner comparable with the way Debye favoured an anti-realist interpretation of

---

[31] Although the *right* way forward in the Bohr case was also very complicated and radical, of course (the new QM).

[32] Norton (1999) provides the details.

certain features of orbits in the Bohr atom. An important difference is that, to avoid inconsistency, in the Bohr case one needs to pretty much remove all of the physical content and embrace instrumentalism, whereas in the Newtonian cosmology case the physical content of the relevant propositions can remain largely intact. I will look at the 'anti-realist re-interpretation' strategy for inconsistencies in more detail in §8.2.3.

What should be especially obvious in the Newtonian cosmology case is that any claim that we need to ask how scientists 'avoided logical explosion by ECQ' is misguided. First, it is clear that for the most part scientists weren't drawing inferences from all of the inconsistent propositions, for any of (I1)–(I4). One reason is that unlike Bohr's theory and CED we just don't have the same sort of relationship between a set of propositions and an explanandum domain of phenomena. But even if scientists *were* making use of (a), (b), and (c) in the eighteenth century, and drawing inferences from these propositions, it is very clear why they didn't infer anything and everything by ECQ. As emphasized in previous chapters, to do so one needs to go *via* a contradiction or explicit contradictories. But nobody ever did reach contradictories, either because they didn't reason with the inconsistent propositions at all, or they didn't reason with them in the right way (because they didn't ask the right question), or because they didn't realise that inference (I★) is truth-preserving.

When it comes to examples from physics such as Newtonian cosmology, much of the inferential work happens within the mathematics. Thus in the light of the present example one very important question is, 'How do we decide whether a given mathematical inference is truth-preserving?' It turns out that we cannot rely on operational rules, as if mathematics were simply an extension of logic: sometimes the legitimacy of an inference depends on subtleties in the *interpretation* of one's mathematics. This will become especially clear in the next chapter, where we turn our attention to the early theories of the calculus courtesy of Newton and Leibniz.

6

The Early Calculus

6.1 Introduction

The example in this chapter will be the early calculus as introduced by both Newton and Leibniz. This has been widely drawn on in the literature to establish that there are many examples of inconsistent theories. Lists of such examples which include the early calculus are to be found in Lakatos (1966: 59), Feyerabend (1978: 158), Priest and Routley (1983: 188), Shapere (1984a: 235), and da Costa and French (2003: 84). Both Newton's 'calculus of fluxions' and Leibniz's 'differential calculus' are referred to. So entrenched is the understanding that the early calculus was inconsistent that many authors don't provide a reference to support the claim, and don't present the set of inconsistent propositions they have in mind. Priest and Routley (1983) do cite a passage from Boyer's esteemed history of the calculus:

[Circa 1720] mathematicians still felt that the calculus must be interpreted in terms of what is intuitively reasonable, rather than of what is logically consistent. (Boyer 1949: 232)

However, Boyer probably isn't referring to an inconsistency in the early calculus at all in this particular passage. What he's saying is that mathematicians in the seventeenth and eighteeenth centuries demanded that mathematics be intuitively reasonable, when they should only have demanded that it be consistent.[1] He's not saying that it wasn't consistent. On the other hand Priest and Routley are probably right that Boyer saw an inconsistency in the early calculus, but then Boyer also doesn't present the set of inconsistent propositions in question, much less explain for us in what way the inconsistency is interesting and/or important.

---

[1] Cf. Boyer (1949: 227): 'Berkeley was unable to appreciate that... its [mathematics'] criterion of truth is inner consistency rather than plausibility in the light of sense perception or intuition.'

What we do find in Boyer and others is a reference to Bishop George Berkeley, who criticized the calculus with his famous 1734 publication *The Analyst*. This is my starting point in §6.2. We are led to an important distinction between the algorithm used to derive results, and the justification of the moves made within the algorithm. The former is discussed in §6.3, and the latter in §6.4. The question of the historical relevance of the relevant propositions takes us on a journey through the commitments of Newton and Leibniz and the mathematical communities in England and France following them. Johann Bernoulli stands out as making the strongest explicit commitment to a set of inconsistent assumptions, but others do at least appear to commit to inconsistency in an 'as if' sense. The latter sense of inconsistency is considered in §6.5; §6.6 is the conclusion.

6.2 Berkeley's Criticism

An example will be useful: imagine (merely for illustrative purposes!) Galileo dropping an apple from the top of the tower of Pisa. He makes measurements to establish that the relationship between the distance travelled s (let's say in metres) and the time taken t (let's say in seconds) is $s = 5t^2$. But he finds it much harder to measure speed, so he wants to calculate this mathematically. What is the apple's speed at the precise moment it hits the ground? Galileo could draw a graph for $s = 5t^2$, then he could find the speed of the apple at any time or distance by finding the slope of the tangent to the curve at that time or distance. It's then obvious how to achieve an excellent approximation to the tangent, and thus the speed.

Consider Figure 6.1. To get an approximation to the tangent at the point (s, t) one can simply take another point on the curve $(s', t+o)$, and draw a line through the two points. If the approximation is not accurate enough we can reduce o, taking the second point as close as we want to the first point to get as good an approximation as we want. This is fine for practical purposes, but of course it isn't much good for achieving an exact result. We need two *different* points to draw a line through, but then we never get the exact tangent at the *single* point (s, t). This is the basic problem. Mathematics is supposed to be an exact science; 'close enough for practical purposes' is not tolerable.

148   UNDERSTANDING INCONSISTENT SCIENCE

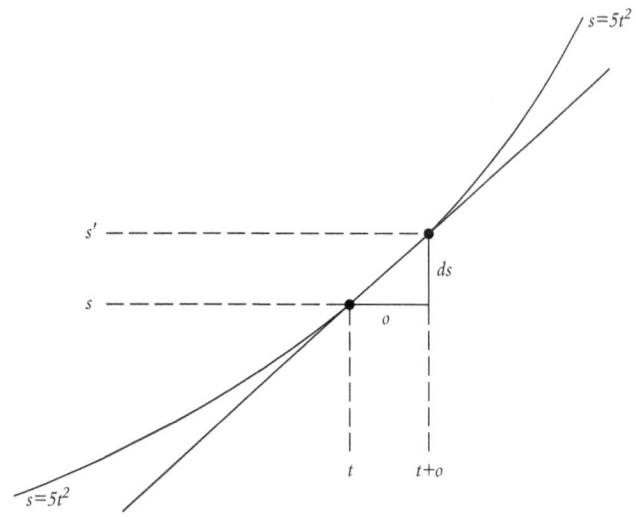

Figure 6.1 An early method for calculating the tangent to a curve. The closer the two points on the curve, the better the approximation to the 'true' tangent at (s, t)

Newton's method, applied to this specific case, followed the latter illustration rather closely.[2] Taking the two points (s, t) and $(s', t+o)$ we start with the expression:

$$s' - s = 5(t + o)^2 - 5t^2$$

Now we expand the brackets and cancel to give:

$$s' - s = 10to + 5o^2$$

Now the slope of the tangent (and thus the speed) is given on the left hand side (LHS) by dividing through by o (change in distance divided by change in time):

$$\frac{s' - s}{o} = 10t + 5o \qquad (6.1)$$

---

[2] Of course, this method was widely used before Newton and Leibniz by the likes of Fermat, Roberval, and Barrow. The principal innovation of Newton and Leibniz was to unify a range of problems and methods for solving those problems, and in particular to see the inverse relationship between tangent and quadrature problems (differentiation and integration). See e.g. Boyer (1949) for the history.

Now we can cancel any remaining terms on the right hand side (RHS) which contain o, and we achieve:

$$\frac{s' - s}{o} = 10t \qquad (6.2)$$

which is of course the right answer.

Clearly the pressing question is, on what grounds do we 'cancel any remaining terms' in the final step, equation (6.1) to (6.2)? The fact that o can be made very small doesn't seem to be good enough, since to cancel these terms it would have to be made equal to zero. But it *cannot* be made equal to zero, since then we would no longer be considering two different points on the line. At the point where we have reached equation (6.1), Berkeley explains the situation as follows (making adjustments for the given example):

Hitherto I have supposed that [t]... hath a real increment, that o is something. And I have proceeded all along on that supposition, without which I should not have been able to have made so much as one single step. From that supposition it is that I get at the increment of [$5t^2$], that I am able to compare it with the increment of [t], and that I find the proportion between the two increments. I now beg leave to make a new supposition contrary to the first, i.e. I will suppose that there is no increment of [t], or that o is nothing; which second supposition destroys my first, and is inconsistent with it, and therefore with every thing that supposeth it. I do nevertheless beg leave to retain [$10t$], which is an expression obtained in virtue of my first supposition, which necessarily presupposeth such supposition, and which could not be obtained without it: All which seems a most inconsistent way of arguing. (Berkeley 1734: *The Analyst*, §XIV)[3]

He later mimics the mathematician as follows:

Let me contradict my self: Let me subvert my own Hypothesis: Let me take it for granted that there is no Increment, at the same time that I retain a Quantity, which I could never have got at but by assuming an Increment. (§XXVII)

Prima facie the charge of inconsistency is clear enough: o is at first necessarily assumed to be *something*, and then later on assumed to be *nothing* in order to

---

[3] Berkeley is actually attacking Newton's 'theory of fluxions' here, but he attacked Leibniz's 'differentials' in just the same way. The difference between fluxions and infinitesimals/differentials will be further explicated in §6.4.

150　UNDERSTANDING INCONSISTENT SCIENCE

cancel terms which include it. However, we must ask whether Berkeley is being fair here, or whether he is attacking a straw man. What exactly is the set of inconsistent propositions in question, and did the relevant individuals commit to those propositions? And what do we learn from the inconsistency given the character of that commitment?

6.3 The Algorithmic Level

Theory eliminativism is once again methodologically helpful here. It would be most painful to ask the question 'What *is* the early calculus?' or 'What is the content of the early calculus?' as a means to answering the question 'Is the early calculus *really* inconsistent or not?' Instead, as seen in previous chapters, we learn everything about inconsistency that it is important to learn by working in terms of sets of propositions. The first question, then, is which are the inconsistent propositions in this case?

Drawing on the discussion in the previous section the obvious propositions concern the properties of *o*, the 'increment' or 'infinitesimal'. Suppose we assume that *o* is a number or quantity, with just as much claim to 'reality' or 'existence' as other numbers. Then suppose we assume that it is non-zero, *and* that it is zero, in order to justify the moves in the derivation given above. Then we will have contradictory commitments.

However, it's clear that questions of justification and assumptions about the properties and metaphysics of the increment *o* were *not* at the forefront of the minds of most mathematicians in the days of the early calculus. Certainly in order to *apply* the techniques of the calculus one did not have to make such assumptions. Indeed, one was encouraged not to. Instead one was urged to follow a set of rules, a procedure, an (informal) algorithm for the relevant class of calculations. The steps are roughly as follows:

1. Put your equation in the form $y = f(x)$.
2. Calculate $\frac{f(x+o) + f(x)}{o}$, and simplify.
3. Remove any terms which are multiples of *o*.

The resulting term is then the derivative.[4]

---

[4] Of course there was more to the calculus than this basic algorithm, but I will focus on this simple account as representative. This is all that will be required to make the important points about inconsist-

Many have called this algorithm 'the calculus'. Edwards (1979) is typical in referring to the early calculus as 'a powerful algorithmic instrument for systematic calculation' (1979: 190). In his well-known annotation of Newton's mathematical works, D. T. Whiteside (on Newton's behalf) entitles one article from c.1665 'The Calculus becomes an Algorithm'. Many others at least make a distinction between the calculus as an algorithm and the justification *of* the calculus (or the moves made within the algorithm). Kitcher (1973: 36–7 and §4) distinguishes between 'the algorithmic level' and the level of justification.[5] Lakatos (1966) also emphasized this distinction, describing it as 'metaphysical versus technical' and 'instrument versus interpretation'. Moving further back still, Baumann made the distinction explicitly in 1869 (p.55), using the term 'the calculus itself' to refer to the algorithm:

Thus we discard the logical and metaphysical justification which Leibniz gave to the calculus, but we decline to touch the calculus itself. (Cited in Lakatos 1966: 58)

Now, the point of all this is certainly not that *we should think of the early calculus as an algorithm*. Theory eliminativism encourages us not to think of 'the calculus' as anything, but rather to present what is most relevant given the history, and to investigate the propositions, equations, etc. from that history in order to learn the important lessons about how science/mathematics works. The point is that in this case, when one looks to the history, the relevant mathematics was largely working as an algorithm. Berkeley himself nearly always referred to the calculus as the 'method of fluxions', and occasionally as the 'algorithm of fluxions' (see e.g. Lavine 1994: 25). Indeed, he noted the distinction I have emphasized, writing 'It appears that his [Newton's] Followers have shewn themselves more eager in applying his Method, than accurate in examining his Principles' (*The Analyst*, §XVII). And Newton's followers had good reason to do just this. Newton and Leibniz (acting independently of course) both adopted an instrumentalist

---

ency. In addition, nothing hangs on my use of the word 'algorithm' to describe this procedure, and the reader who objects to this terminology should simply read 'informal algorithm' for 'algorithm' in what follows. 'Informal algorithm' is hardly a contradiction in terms, given the complications surrounding the concept *algorithm* (Cleland 2001, 2002). At any rate, it is usual in the relevant literature to use the word 'algorithm', as we will see shortly.

[5] What amounts to the same distinction is found throughout the literature, including Bos (1980: 60), Grabiner (1983: 16), Guicciardini (1989: 38), Jesseph (1993: 148–50), and Arthur (1995).

attitude in the majority of their publications, with only marginal comments on explanation or justification. As Kitcher (1973) writes,

> Newton typically expressed his algorithms in the form of a set of instructions to the reader. The mechanical application of his rules to the problems at hand would then yield their solutions. (Kitcher 1973: 37)[6]

The first sign of Newton's calculus in print was the *Principia* (1687). As Boyer writes:

> His contribution was that of facilitating the operations, rather than of clarifying the conceptions. As Newton himself admitted in this work [*Principia*], his method is 'shortly explained, rather than accurately demonstrated.' ... [M]athematics was a method rather than an explanation. (Boyer 1949: 193)

Guicciardini (2004) writes as follows:

> Newton did not [in the *Principia*] give the reader any detail on these highly algorithmic techniques. When David Gregory and Roger Cotes approached their master in order to have details on how to complete certain proofs which depend upon 'the quadrature of curves', Newton revealed to his disciples how to apply his integral tables to several problems faced in the *Principia* via letters or oral communications. (Guicciardini 2004: 465f.)

His *De Quadratura* (1704) had a similarly instrumental flavour (see §6.4.1, below). And his *De Analysi* (1711) begins by stating the methodological rules of the calculus, without justification, before going on to present various examples (Edwards 1979: 201).

Turning to Leibniz we find a similar story. Mancosu (1996) writes of Leibniz's first publication of the calculus, the *Nova Methodus* of 1684,

> The paper is remarkable for the paucity of the explanations given by Leibniz.... Leibniz does not explain how he arrived at his equations and leaves the reader totally in the dark as to the heuristics and formal proofs of the results therein presented. (Mancosu 1996: 151)

---

[6] See Kitcher's (1973) paper for some examples taken directly from the primary evidence, and references to further examples. In fact both the primary and secondary literature is jam-packed with examples of algorithms presented as sets of instructions, by Newton, Leibniz, and also by Newton and Leibniz's followers.

He continues (1996: 153), 'In 1684 Leibniz has presented without justification the . . . rules for the calculus' (cf. Bos 1974: 63). And Arthur (2008) notes that this is not unique to the *Nova Methodus*:

> Leibniz often stressed the pragmatic utility of his techniques, and how they could be exploited by mathematicians without their having to trouble themselves with foundational problems. (Arthur 2008: 20)

Of course both of the founders did spend *some* time attempting to justify their procedures (as we'll see in the next section). But it is important to realize that these 'justifications' were consciously distinguished from the procedures at the time. And not just by the founders but by the communities following them. As Kitcher writes,

> The mathematical community had appreciated the power of Newtonian and Leibnizian techniques, and had shelved worries about the explanation of their success. (Kitcher 1983: 256)[7]

We'll see more evidence of this below, when I consider the attitudes towards justification of the English and French communities. Of course, there were *some* objectors (as we've already seen with Berkeley), but the success of the techniques was such that the vast majority didn't worry about the justification of the moves made within the algorithm. This was possible in large part because of how revolutionary the calculus was at that time. Before Newton and Leibniz there had been just a handful of disparate techniques for solving what were really calculus problems. Now, with the new algorithms, a multitude of previously intractable problems were made accessible, and the calculus produced success after success. In this environment the founders were able to present the calculus without explaining *why* it worked.

Clearly, then, the calculus conceived as an (informal) algorithm has a great deal of what I have been calling 'historical relevance'. It was both used and conceived of as an algorithm at the time, at least for the most part, by many of the relevant individuals, including the two founders. Is an algorithm even the *kind* of thing which can be inconsistent? So far in this book I have focused on a broad notion of 'propositions' (including sentences,

---

[7] Cf. Bos (1980: 80): 'Most mathematicians spend most of their time not in contemplating these concepts and methods, but in using them to solve problems.'

statements, assumptions), because these are the kind of thing which obviously can be consistent or inconsistent. An algorithm on the other hand is not a set of propositions, but instead an ordered list of directions to the user.

There are three reasons we might want to call an algorithm 'inconsistent'. First we can get a sort of inconsistency in an algorithm akin to the liar paradox:

(1) Do not carry out this instruction.

Clearly the algorithm of the early calculus does not have this problem! More seriously, we might refer to an algorithm as 'inconsistent' if it gives as outputs any logical absurdities such as '2 = 1'. An assumption all users *were* making is that the algorithm gave correct results. In other words, they considered the following assumption a (very) serious candidate for the truth:

If, for any function $y = f(x)$, we calculate $\frac{f(x+o)-f(x)}{o}$, where o acts as a numerical constant, and then delete any remaining terms which are multiples of o, then we are left with the derivative.

And they obviously considered the other basic assumptions of mathematics as (very) serious candidates for the truth. But then it does seem possible that an important inconsistency could arise here just if the algorithm gives a single false result. As an example consider the following equality:

$$\frac{\pi}{4} = 1 - \frac{1}{3} + \frac{1}{5} - \frac{1}{7} + \frac{1}{9} - \ldots$$

This equation was derived by Leibniz with the use of his calculus. Now if the calculus had led to something slightly different, so that it followed that $\pi > 4$, then we would have a contradiction since by definition $\sim\pi > 4$. This would be quite a serious inconsistency, comparable to the Pauli inconsistency and the (I2) inconsistency from Newtonian cosmology, in the sense that all of the propositions were considered serious candidates for the truth.

But in fact the early calculus was not inconsistent in this sense: it never did give a wrong result.[8] Certainly Berkeley was happy to admit this, writing

---

[8] At least not in the form I have presented it. More exotic applications led to false results in the hands of Cauchy and others in the nineteenth century. These results have never been mentioned as responsible for the inconsistency of the early calculus!

that 'Analysts arrive at truth'. His objection was rather that they didn't know *how* they arrived at truth (§XXII). Many no doubt saw this as overly charitable, since some *did* see falsehoods as following from the early calculus. For example, consider the following calculation:

(i) $x + o = x$
(ii) $x + 2o = x + o$
(iii) $2o = o$
(iv) $2 = 1$

Equation (i) seems to follow from the calculus, since it embodies the final step of the procedure where one drops all terms which are multiples of o (in the example in §6.2, recall, the final step suggests that $10t + 5o = 10t$). Equations (ii), (iii), and (iv) then follow from natural application of the algebraic laws, and one reaches the contradiction $2 = 1$. Has this then followed *from the calculus*? It certainly hasn't followed from the algorithms accepted by the relevant communities at the time. The algorithms only permit one to drop terms multiplied by o in the final stage of the specified procedure. This was widely acknowledged at the time, even by most of the critics; Berkeley was certainly willing to accept this, since he made no such objection.

Others did try to make something of this. Michel Rolle, in the early 1700s in France, claimed to be able to prove that $o = 0$, and also that the calculus led to mistakes. However, the calculation leading to $o = 0$ stepped outside the accepted algorithm, as with the example just given above, and the alleged mistakes were shown to follow from mistaken application (see §6.4.4, this volume). The calculus stood up to the challenge in every case; no mathematical falsehoods followed from the application of the algorithm, correctly understood.

This discussion also rules out a third possible sense in which we might refer to an algorithm as 'inconsistent': if it allows for different, contrary outputs given the same input. If two mathematicians asking the same question achieved contrary answers, and both had faithfully followed the rules of the procedure and hadn't made any mistake, then we might reasonably call the algorithm in question 'inconsistent', and indeed we probably wouldn't want to call it an 'algorithm' at all. But given that

'Analysts arrive at truth', it is clear that the early calculus did not lead to different, contrary results.[9]

6.4 The Level of Justification

The reader may insist that the calculus *as an algorithm*, as presented above, obviously isn't inconsistent and was never the target of the inconsistency claim. For the benefit of such a reader I must emphasize that I don't have an agenda here, as if I am looking to deflate all of the inconsistency claims in the literature, perhaps because I want to argue that science is more rational than it is sometimes said to be. There is no such agenda. The goal is merely to examine the inconsistency claims which have been made in the light of the true complexity of science and the history of science. The algorithm discussed in the previous section is not supposed to be 'what the calculus is'—I am making no claim at all about 'what the calculus is'. The algorithm is presented because it is highly relevant to the mathematical activities under discussion in this chapter. It should already give those who believe 'the early calculus was inconsistent' pause for thought if much of the mathematical work was done not with a set of inconsistent propositions, but with an algorithm, a set of rules to be carried out in a specified order, and such that the rules never led to two contrary outputs for the same input.

But let me now turn to what, for many, is no doubt the inconsistency of real importance here. I have already mentioned it above, as an inconsistency in the properties of the 'increment' or 'infinitesimal' o. It seems on the face of it that these inconsistent properties are necessary to justify the moves made within the algorithm. To be explicit, the assumptions appear to be as follows:

(1) o is a quantity (the same kind of thing as a natural number).
(2) o is non-zero (at the beginning of one of the relevant calculations).
(3) o is zero (in the final step of the calculation).

---

[9] Herein lies a further justification for using the word 'algorithm'. The given procedure has in common with an algorithm (but *not* a recipe, at least in my kitchen!) the feature that it reliably led different mathematicians from same-input to same-output.

This is perhaps the most obvious explanation of the moves made within the algorithm. It is also completely unsatisfactory, precisely because of the blatant inconsistency, and one may wonder if anybody really took it seriously and realistically as a literal explanation. The inconsistency is extremely obvious: there is no question of it going unnoticed as in the later Bohr theory or (I2) in Newtonian cosmology. There also seems to be no possibility of it being tolerated if the propositions are committed to as candidates for the truth. Recall that when the inconsistencies in the Pauli case and Newtonian cosmology case were eventually noticed, it was interpreted as a crisis demanding immediate and urgent response, not something that could be tolerated. Any commitment to the propositions as candidates for the truth was immediately suspended.

It will be demonstrated in the following subsections that (as expected) the propositions (1)–(3) were not committed to as candidates for the truth. But then what exactly was their role? And if thinkers at the time didn't commit to them literally as justifications for the moves made in the algorithm, then what justification *did* they have? We can say with some confidence that nobody at the time had a suitable justification: this only came much later with Cauchy and others in the nineteenth century. One option is then clear: individuals might have had *no justification at all*. It was perfectly possible to use the algorithm, and to feel justified in using it (given successful results), without being able to give any justification for the individual moves made *within* the algorithm. Indeed, such an attitude has been advocated on numerous occasions in the history of science and mathematics. For example Oliver Heaviside wrote in 1899,

Shall I refuse my dinner because I do not fully understand the processes of digestion? No, not if I am satisfied with the result.... First, get on, in any way possible, and let the logic be left for later work. (Heaviside 1899: 9)

It should be no surprise if we find such an attitude in many of those who used the early calculus.

A second option is that individuals *did* attempt an explanation, but that it is inadequate. This would be a proposed explanation, perfectly consistent and coherent, but which in fact on examination doesn't explain the moves made within the algorithm. Finally, some may have given explanations which were simply incoherent: inadequate because metaphysically or conceptually unclear.

In this section I will be on the lookout for these three possibilities as sensible alternatives to the usual story we find in the literature. When we look to the relevant individuals and communities with whom we associate the early calculus, did they commit, in one way or another, to the blatantly inconsistent explanation noted above? Or did they rather have (a) no explanation at all, (b) an inadequate explanation, or (c) an incoherent explanation?

6.4.1 *Newton*

In the previous section I emphasized that Newton usually presented the calculus in the form of algorithms, without any concomitant justification. But it's true that there are important passages where Newton does discuss the question of the justification of the moves made within the algorithms. *Prima facie* it looks like Newton presents three different justifications, one in terms of 'infinitesimals', another in terms of 'fluxions', and a third in terms of 'primary and ultimate ratios' (Boyer 1949: 219).[10]

Of course in the search for inconsistency our attention turns to the infinitesimal justifications. Boyer for one has written that,

Berkeley's objection to Newton's infinitesimal conceptions as self-contradictory was quite pertinent. (Boyer 1949: 226)

However, today there is a consensus in the literature that Newton's explanation of the calculus was much more sophisticated than this would suggest. Even from the very beginning, before Newton had distributed anything to his contemporaries, there is clear evidence that he had reservations about basing the calculus on the notion of infinitesimals. Guicciardini (2004) writes as follows:

While he was adamantly sure about the validity of his mathematical results, Newton felt acutely that the status of infinitesimals was controversial. He was not a man who could easily accept the uncertain status of infinitesimals by appealing to their heuristic power. (Guicciardini 2004: 463; see also Arthur 1995: 340)

---

[10] There is a vast literature on Newton and the calculus, to which I cannot hope to do justice here. As I mentioned in Chapter 2, this is not history of science, but historically informed philosophy of science. The amount of historical awareness necessary to decide what role inconsistency played in the early calculus is not great, by historians' standards, but certainly more than is usual in the philosophy of science literature.

Indeed, Newton's reservations about infinitesimals are explicit enough in the primary literature, and are now well documented. Clearly his first priority in the 1660s was to develop techniques mathematicians could use to solve problems, leaving the justification of those techniques for another day (just as Heaviside would recommend). As Kitcher (1973) puts it,

> [T]he existing methods were reckoned unsatisfactory either on grounds of inconvenience or lack of generality—in most cases on both counts—and, aside from questions of justification, there was evident need at the algorithmic level for extension and simplification of the methods used. Newton began by accepting the algorithmic challenge. (Kitcher 1973: 36)[11]

It is also accepted knowledge that Newton held back from publishing his work on the calculus as long as he did in large part because he didn't yet have a justificatory account that he deemed satisfactory. As Guicciardini (2004) puts it,

> [F]rom the 1670s, Newton realised that the standards of validation that he aimed at were often above his mathematical practice, a practice which he was led to downgrade to the level of a heuristic technique not worthy of printed publication. (Guicciardini 2004: 459f.)[12]

This given, the suggestion that he committed to propositions (1)–(3) is just mistaken (in any obvious sense of 'committed to'; cf. §6.5, below). There is good evidence that, for a while, he didn't quite know what he believed vis-à-vis foundations: he slowly developed his justificational account via the concepts of 'fluxions' and 'prime and ultimate ratios'. When he did finally publish on these issues it was in Book 1 of the *Principia* in 1687, more than 20 years after his 'discovery' of the calculus, and 18 years after he wrote (but did not publish) his first major work on the calculus *De Analysi* (1669).

Before I turn to Newton's own words in the *Principia*, it is important that I clarify Newton's terminology, especially what he meant by a 'fluxion'. The fact was that Newton and others at that time were brought up on a kinematic conception of lines and curves: one was to imagine a curve, for

---

[11] Kitcher (1973: 43): 'At an early stage of his career Newton gave several hints that he was not satisfied to let justification rest with the method of infinitesimals. It was obviously natural for him to ignore these worries until analysis had reached its goal.... [A]nalysis and the development of algorithms were his main concerns.'

[12] See also p.464: '[In the 1670s] Newton came to the conclusion that the "new analysis" was not "worthy to be read": it was not a mathematical language fit for publication.'

example represented by $y = 5x^2$, as generated by a moving point (see e.g. Arthur 1995: 338). In thus generating a curved line there would then be two dimensions of movement: in the 'x' and 'y' directions in the case of $y = 5x^2$. These (x and y) Newton called the 'fluents'. Since we are imagining movement along the line, there is then the question of the velocity in each of the x and y directions at any given point in time: these are the 'fluxions', represented by $\dot{x}$ and $\dot{y}$. If one then considers a very small increment of time o one can consider how much each fluent will have 'flowed' in that time.[13] It will be $\dot{x}o$ and $\dot{y}o$, which Newton calls 'moments' (although they are not moments in time, but instead distances of 'flow' of the fluents x and y).

In practice this helps us reach the slope of the tangent at a given point (x, y) as follows.[14] The slope will be given by the ratio of the fluxions—$\dot{y}/\dot{x}$—at the point (x, y).[15] We consider the flow of the fluents $\dot{x}o$ and $\dot{y}o$ in an increment of time o, and in so doing reach a new point on the curve ($x + \dot{x}o$, $y + \dot{y}o$). We can draw on the relationship between x and y ($y = 5x^2$) to write down the relationship between $x + \dot{x}o$ and $y + \dot{y}o$:

$$y + \dot{y}o = 5(x + \dot{x}o)^2 = 5x^2 + 10x\dot{x}o + 5\dot{x}^2o^2$$

Now we know that $y = 5x^2$, so we can exchange the 'y' (at the front) accordingly, and then we can simplify and we reach:

$$\dot{y} = 10x\dot{x} + 5\dot{x}^2 o$$

Now one eliminates the final term $5\dot{x}^2 o$, and one reaches the conclusion $\dot{y}/\dot{x} = 10x$.

This derivation can be compared with what we saw in §6.2, above. It would seem that there is no substantial difference, except that instead of an infinitesimal distance o, one makes use of an infinitesimal amount of time o and a corresponding infinitesimal flow of the fluents: $\dot{x}o$ and $\dot{y}o$. One still has the problem of justifying the elimination of terms involving 'o' in the final

---

[13] Here the flow of time refers to the time taken to generate the curve; I've changed the example from $s = 5t^2$ to $y = 5x^2$ to emphasize this: 'x' certainly does not have to represent 'time'.

[14] An example Newton gave concerned finding the tangent to the curve given by the equation $x^3 - ax^2 + axy - y^3 = 0$. Kitcher (1973: 43) draws from the primary source, but a simpler example will suffice for present purposes.

[15] To make sense of this it is possible to think of $\dot{y}/\dot{x}$ as $\frac{dy}{dt}\frac{dt}{dx} = \frac{dy}{dx}$, although Newton certainly wouldn't have used this Leibnizian symbolism.

step. And this was certainly not lost on Berkeley. On the other hand, one might invoke a fluxional analysis but *not* appeal to an 'infinitesimal' increment of time. Instead one might hint at a limiting procedure, perhaps appealing to intuitions about the continuity of the flow of time. (There is some disagreement in the literature on this issue: cf. Kitcher 1973: 40 and Arthur 1995: 343.)

I will return to fluxions in §6.4.3, when I consider the attitudes of the English community following Newton. For now it is enough to note that Newton himself certainly didn't consider the ultimate justification of his algorithms to be based on infinitesimals, or solely on kinematic intuitions about the continuity of the flow of time. On the contrary, he developed his fluxional analysis in 1665 (see e.g. Whiteside and Newton 1967: 146), and yet still held back from publication for over 20 years because (or *mainly* because) he didn't consider this a proper justification. In fact, Kitcher (1973) has argued that fluxions are not justificational at all, but instead 'yield the heuristic methods of the calculus' (1973: 33f.).[16] Whatever the case, when Newton finally *did* publish, he had developed a quite different justification involving 'prime and ultimate ratios'.

Newton's own words—as presented for example on in Section 1 of Book 1 of the *Principia*—are Newton's clearest justificational message.[17] There he includes a number of 'lemmas', including,

Quantities, and the ratios of quantities, which in any finite time converge continually to equality, and before the end of that time approach nearer to each other than by any given difference, become ultimately equal. (Newton 1729 [1687]: 41f.)

What does Newton mean by quantities 'converging continually to equality'? The idea is to consider the '*o*' as an increment of time, as in the example above, and to consider it getting increasingly smaller, for example because it is being halved again and again. Then we should consider what happens to

---

[16] Arthur (1995) argues that Newton's fluxions do play a role in justification, at least insofar as they act to eliminate infinitesimals. This is achieved primarily through the conception of time as truly continuous: '[O]n the hypothesis of a continually flowing time, there can be no question of allowing indivisibles. And indeed, once Newton has invented his fluxional geometry, he never refers to moments as indivisible. In the *Principia* (as in all the later justifications Newton gives of his Method of Fluxions), he makes a point of insisting on this' (1995: 342).

[17] Here I draw on Brook (1973: 184–5). But these passages are quoted in much of the relevant literature to show how sophisticated, and relatively modern in character, Newton's position really was (e.g. Kitcher 1973: 47–8; Edwards 1979: 225; Arthur 2008: 10ff.).

Table 6.1 Considering again the example introduced in §6.2, as we take the value of 'o' to get smaller and smaller as it is halved again and again, the slope of the tangent gets closer and closer to 30. There are no 'ultimate quantities' at the bottom of this list; instead one conceives of it as going on forever, '*in infinitum*'.

Size of 'o'	Slope of tangent for $s = 45\text{m}$
1	35
1/2	32.5
1/4	31.25
1/8	30.625
1/16	30.313
1/32	30.156
1/64	30.078
1/128	30.039
1/256	30.020
1/512	30.010
...	...

the slope of the tangent. In the case represented in Table 6.1, the slope we derive from our calculation gets continually closer to 30 and so, Newton tells us, it ultimately *becomes equal to* 30. But how is this possible if o gets smaller and smaller, without ever actually reaching zero?

Newton presents several 'lemmas'. He then clarifies as follows:

These lemmas are premised to avoid the tediousness of deducing involved demonstrations ad absurdum, according to the method of the ancient geometers. For demonstrations are shorter by the method of indivisibles; but because the hypothesis of indivisibles seems somewhat harsh, and therefore that method is reckoned less geometrical, I chose rather to reduce the demonstrations of the following propositions to the first and last sums and ratios of nascent and evanescent quantities, that is to the limits of those sums and ratios, and so to premise, as short as I could, the demonstration of those limits. For hereby the same thing is perform'd as by the method of indivisibles; and now those principles being demonstrated, we may use them with more safety. Therefore if hereafter I should happen to consider quantities as made up of particles... I would not be understood to mean indivisibles, but evanescent divisible quantities; not the sums and ratios of determinate parts, but always the limits of sums and ratios.... It may also be objected, that if ultimate ratios of evanescent quantities are given, their ultimate magnitudes will also be given: and so all quantities will consist of indivisibles, which is contrary to what Euclid has demonstrated concerning incommensurables, in the tenth book of his elements. But this objection is founded on a false supposition. For those ultimate ratios with which quantities vanish are not truly the ratio of ultimate quantities, but limits towards which the ratios of quantities decreasing without limit do always converge; and to which they approach nearer than by any given difference, but never go beyond, nor in effect attain to, till the quantities are diminished *in infinitum*. (Newton 1729 [1687]: 54f.)

Of course there is some difficult terminology here, but to attempt a full interpretation of Newton would take us too far astray. Instead I present Newton's own words to give the reader a sense of the primary literature. Drawing on such passages (including another passage from the introduction to Newton's *De Quadratura*, published in 1704) Edwards writes,

In other words, Newton says, exposition in terms of indivisibles or infinitesimals is simply a convenient shorthand (but not a substitute) for rigorous mathematical proof in terms of ultimate ratios (limits). (Edwards 1979: 226)

And any other Newton scholar who has worked in this area comes to a similar conclusion:

The common answer of the fluxionists [Newton's followers] was that Berkeley's logical criticism was applicable only to the differential method, which was employed by Newton merely to abbreviate the proofs. Newton's genuine method was the method of limits. (Guicciardini 1989: 44)

The theory of infinitesimals was to abbreviate the rigorous proof [in terms of ultimate ratios], and Newton thought he had shown the abbreviation to be permissible. (Kitcher 1973: 34)

[The criticisms were] uncharitable. Berkeley's reading presupposes the unfavourable interpretation of the argument. (Kitcher 1983: 239, fn.15)

[Berkeley's] criticism may be a bit unfair to Newton, who can, as we have seen, be read as having some idea of using something like limits to replace the procedure of setting *o* equal to zero. (Lavine 1994: 24)

Now this isn't to say that Newton had a perfectly good justification of his algorithm, and that this really should have been noticed by Berkeley and others. Exactly what we should say on this is still a point of some dispute in the literature. Is Newton saying that the infinite series never actually reaches the 'ultimate ratio' or 'limit'? And if not, then why should we agree that it ultimately 'becomes equal'? Responding to Kitcher (1983) and Sherry (1987), Jesseph (1993) warns us not to be *too* charitable to Newton here:

[I]t is of no use to show that charitable interpretations of Newton can evade Berkeley's critique if such interpretations were not part of the mathematical landscape of the 1730s. (Jesseph 1993: 198)[18]

However, this debate is irrelevant to the point I am making here: that Newton's justification of the moves made in his algorithm was not inconsistent. Suppose Newton's justification ultimately *does* fail to justify the practice of eliminating terms containing o in the final step of the derivation. Then Newton's justification isn't adequate: it doesn't do the job he intends it to do, and his talk of 'infinitesimals' or 'indivisibles' should be taken as shorthand for this inadequate justification. This would still make it inappropriate and misleading to talk in terms of *inconsistent* commitments here.

Newton's commitment to a 'rigorous' justification in terms of 'limits', and his rejection of 'infinitesimals' or 'indivisibles', continued throughout his career. As in the *Principia*, in his second major publication on the calculus *De Quadratura* (1704) a disclaimer tells us not to take talk of infinitesimals literally.[19] However, after 1700 Newton did in fact start to publish some of his earliest material, wherein one finds more talk of infinitesimals. For example, in 1711 he published his *De Analysi* which he had actually written more than 40 years earlier, in 1669. The main reason for these publications after 1700 was that Newton had become increasingly aware of not only Leibniz, but also other mathematicians such as David Gregory trying to claim the discovery of results he had achieved years earlier (see Guicciardini 2004: 465). But as Guicciardini (2004: 466) has demonstrated, before he published his early papers 'He changed his original manuscripts, trying to avoid reference to infinitesimals.'

Enough on Newton. It is clear that he didn't commit to the blatantly inconsistent propositions (1)–(3) noted above. And even if Newton often

---

[18] Brook (1973) argues that Berkeley was unfair to Newton, on the grounds that Newton didn't think there really was an 'ultimate ratio'. Instead, what Newton calls the 'ultimate ratio' is, according to Brook, 'the numerical limit to which an actual sequence of ratios converges. In this sense, although Newton is not explicit on this point, the "ultimate ratio" is not a member of the set of actual ratios (actual comparison of magnitudes) for which it is the numerical limit' (1973: 185). Brook also mentions that other passages seem to go against this, and Newton ultimately isn't quite clear, and consistent (stable in his views), enough.

[19] See Edwards (1979: 226). Cf. Guicciardini (1989: 42): '"De Quadratura" (1704c) was an attempt to base the calculus solely upon limits. But generally Newton's effort was not understood', and Kitcher (1973: 42f.): 'It is unclear whether there is any evidence of Newton asking himself what o denotes (ie, what an infinitesimal is). Indeed, in the light of his *De Quadratura* with its instrumentalist attitude toward infinitesimals, the question would seem to be meaningless for him.'

reasoned *as if* he committed to (1)–(3), there can be no danger of ECQ so long as he derived results using his algorithms (I'll return to 'as if' commitment in §6.5). So it isn't at all clear that there is any good point to describing Newton as inconsistently committed here. However, Newton is just one man: we may wonder whether others at the time were inconsistently committed.[20]

6.4.2 *Leibniz*

Some readers may suppose that Leibniz is a better candidate for real commitment to infinitesimals, together with their contradictory properties. The historical facts belie this supposition. From the age of 30 until his death at the age of 72 in 1716 Leibniz consistently denied the 'existence' of infinitesimals, and on at least one occasion explicitly attempted to prove that infinitesimals were impossible (see Bassler 2008). He repeatedly referred to infinitesimals as 'well founded fictions', which can stand in for full, rigorous derivations of results which do not refer to infinitely small quantities at all. And indeed, in recent decades it has become clear that Leibniz's best attempt at a rigorous foundation for the calculus was at least as good as Newton's. But to address the doubts of the sceptical reader, let me back up and provide evidence for these claims.

Now Leibniz did not reject infinitesimals from Day One. In his early years of mathematical activity, 1669–74, he was arguably a 'realist' about infinitesimals, in some sense (Arthur 2009). However, in the course of writing what is now considered his major early work on the calculus, infinitesimals came to be explicitly rejected. As Levey (2008) puts it,

[W]ithin a few short weeks [in 1676], it's all over for the infinitely small. Leibniz begins confidently describing infinitesimals and their ilk as 'fictions'. (Levey 2008: 114f.)

The 'major early work' I am referring to is his *De Quadratura Arithmetica* (DQA for short), written 1675–76, but only very recently published for the

---

[20] Colyvan (2009) suggests that Newton was inconsistent because he claimed that $x = x + o$, but the LHS of the equation is a constant and the RHS is a variable. Needless to say, this doesn't do Newton justice! Although Colyvan may be more interested in 'rational reconstruction' than historical accuracy. See later in the chapter for more on these issues.

first time in 1993 (see e.g. Knobloch 2002; Arthur 2008; Levey 2008). Just as Newton tells us in the *Principia* that the 'method of indivisibles' is a well-founded, convenient shorthand to avoid the 'tediousness' of a rigorous demonstration, in his DQA Leibniz labels infinitesimals 'fictions' which 'allow economies of speech and thought in discovery as well as in demonstration' (cited in Arthur 2008: 27). In place of Newton's 'first and last ratios' or 'limits' Leibniz presents what amounts to an almost identical 'rigorous foundation' for the calculus. To put it in terms of the example of $y = 5x^2$ of the previous section, one is to consider the fact that, for *any* possible given quantity ϵ, however small, the difference between the exact tangent and the calculated tangent can be made smaller than ϵ by making o small enough (since there is no limit on how small o can be made). In other words the 'error' is smaller than any possible quantity ϵ. But the only thing smaller than any possible quantity is zero. So the error is zero, and we have not inequality, but equality. As Leibniz puts it in DQA, 'the error is smaller than any assignable, and therefore null' (Levey 2008: 117). As Arthur (2008) comments, Leibniz makes,

> ... an inference from the fact that a difference between two quantities can be made smaller than any that can be assigned, to their difference being null. This is a *reductio* in the sense that whatever minimum difference one supposes there to be, one can prove that the difference is smaller. (Arthur 2008: 25)

As Levey (2008: 113ff.) puts it, here Leibniz provides a 'new principle of equality', which can be formalized as,

If, for any n, $|x - y| < \frac{1}{n}$, then $x = y$.

Leibniz did want to publish the DQA, but it didn't get published. It was some time after 1676 when Leibniz introduced his differential calculus to the world in his article *Nova Methodus* published in the *Acta Eruditorum* in 1684. Having to his mind established that infinitesimals were 'well founded fictions' that could be used to abbreviate proofs, Leibniz originally drafted his *Nova Methodus* in the terminology of infinitesimals. However, he had not yet published on his fictionalism, and without a discussion of his foundations in the public realm an article presenting the calculus in terms of infinitesimals was bound to be interpreted as making a 'realist' commitment to infinitesimals. Perhaps to avoid such an interpretation, Leibniz revised the *Nova Methodus* removing all reference to infinitesimals before publication (cf. Horvath 1986: 62).[21]

His reservations about infinitesimals remained in 1689 when he corresponded with Johann Bernoulli (whom we will meet in §6.4.4):

> For if we suppose that there actually exist the segments on the line that are to be designated by ½, ¼, ⅛, ..., and that all the members of this sequence actually exist, you conclude from this that an infinitely small member must also exist. In my opinion, however, the assumption implies nothing but the existence of any finite fraction of arbitrary smallness. (Cited in Rescher 1955: 113)

Then in a letter to Varignon in 1702 he states clearly that, in his view, one doesn't need 'real' infinitesimals to justify the calculus in a rigorous sense, but also that one can use them 'with confidence' as a way to shorten the reasoning:

> These incomparable quantities are not at all fixed or determined but can be taken as small as we wish in our geometrical reasoning and so have the effect of the infinitely small in the rigorous sense. If any opponent tries to contradict this proposition, it follows from our calculus that the error will be less than any possible assignable error, since it is in our power to take that incomparably small quantity small enough for that purpose, inasmuch as we can always take a quantity as small as we wish. May be that is what you mean, Sir, by the notion when you speak about that of inexhaustion, moreover, there is no doubt, that this idea establishes the rigorous demonstration of the infinitesimal calculus.... Moreover, even if someone refuses to admit infinite and infinitely small lines in a rigorous metaphysical sense and as real things, he can still use them with confidence as ideal concepts which shorten the reasoning. (In Horvath 1986: 66)

And in a 1706 letter to Bartholomeus Des Bosses Leibniz is explicit on both his fictionalism about infinitesimals, and his preferred justification in terms of finite quantities:

> Philosophically speaking, I hold that there are no more infinitely small magnitudes than infinitely large ones, i.e. that there are no more infinitesimals than infinituples. For I hold both to be fictions of the mind due to an abbreviated manner of speaking, fitting for calculation, as are also imaginary roots in algebra. Meanwhile I have demonstrated that these expressions have a great utility for abbreviating

---

[21] Cf. this with Newton: I mentioned above that Newton *also* removed talk of infinitesimals from some of his works when it came to publication (Guicciardini 2004: 466).

thought and thus for discovery, and cannot lead to error, since it suffices to substitute for the infinitely small something as small as one wishes, so that the error is smaller than any given, whence it follows that there can be no error. R. P. Gouye, who objected, seems to me not to have understood adequately. (In Levey 2008: 128)

The R. P. Gouye in question was just one of several individuals who objected to the use of infinitesimals over the course of Leibniz's lifetime, bringing Leibniz to repeat his foundational beliefs on numerous occasions. Many statements are in letters such as those referenced above, but some are public. When the Dutchman Bernard Nieuwentijt publicly objected to infinitesimals in 1694 and 1695, Leibniz replied in an article published in the *Acta Eruditorum* in 1695. His response isn't quite as clear as it could be (see Nagel 2008: 203): without a doubt, Leibniz's foundational comments are sometimes unclear and frustrating, or even inconsistent *in the sense that he says different things at different times*. At times it is clear that he is more interested in using the calculus to achieve results than in foundational worries: for example, in his reply to Nieuwentijt he writes at one point 'It is sufficient that [a definition] is intelligible and useful for making discoveries' (Nagel 2008: 203). Leibniz does better in an article published in the *Acta Eruditorum* in 1712:

Just as I have denied the reality of a ratio, one of whose terms is less than zero, I equally deny that there is properly speaking an infinite number, or an infinitely small number, or any infinite line or a line infinitely small. . . . The infinite, whether continuous or discrete, is not properly a unity, nor a whole, nor a quantity, and when by analogy we use it in this sense, it is a certain *façon de parler*. . . I once established that when we call an error infinitely small, we wish only to say an error less than any given, and thus nothing in reality. (In Jesseph 2008: 231f.)

Quite how impressed we should be with Leibniz's foundations is still a matter of some debate. Many scholars are increasingly impressed, especially since the publication of the DQA. Arthur (2008) thinks that Leibniz, in DQA, provides a method which is 'by all relevant standards completely rigorous, being effectively equivalent to what is now known as Riemannian Integration' (2008: 29).[22] But for present purposes I don't need to go this

---

[22] In fact Arthur (2008: 25) argues that Newton's and Leibniz's rigorous foundational stories are fundamentally the same (considering Newton's *Principia* and Leibniz's DQA). He concludes, '[W]e have

far; it is enough if I have demonstrated that Leibniz (as Newton) did not have inconsistent commitments here. Even if Leibniz's foundational account ultimately isn't successful, it remains true that, as Levey puts it,

> [E]xpressions for infinitesimals ... are not designating terms for infinitely small quantities but rather they are shorthand devices for complex expressions that refer only to finite quantities. (Levey 2008: 124)[23]

In such circumstances it may *look* like Leibniz has inconsistent commitments when he really doesn't at all. Jesseph (2008) writes,

> In many contexts Leibniz's account of his *calculus differentialis* is phrased in terms that are most readily interpreted as implying the real existence of infinitely small magnitudes. (Jesseph 2008: 215)

But given Leibniz's clear statements about how we should interpret his infinitesimal-talk, the interpretation which 'most readily' comes to hand is not a fair interpretation at all. Indeed it is an irresponsible interpretation.

We see at this point that it is most inappropriate and misleading to describe either of the two founders of the calculus as committed to an inconsistent theory. However, so far it remains possible that there were wider communities who *did* have inconsistent commitments here. It is to the two most relevant mathematical communities I now turn: the English and the French.

6.4.3 *The English*

Did the English community commit, in some significant sense, to an inconsistent set of propositions here? One obvious reason to say 'no' would be if relevant individuals just didn't think about foundations at all (a 'shut up and calculate' approach). In the early years, before Berkeley's *Analyst* (1734), this seems to be a fair representation. Newton had presented an algorithm that could be applied to get interesting results, and to do calculations which had been very difficult or impossible previously. Developments

---

seen a consilience in the foundational writings of Newton and Leibniz that is quite remarkable...' (2008: 29).

[23] Leibniz does expand on 'fictionalism' in other contexts. For example, in a letter from 1706 he describes 'flocks' of sheep as well founded fictions: although we *talk* as if there is a 'flock' of sheep, in reality there are only the individual sheep staying close to each other and following each other around. See Jesseph (1998: 33ff).

were 'almost automatic', as Boyer (1949: 243) has put it: one could take any of a vast number of problems and apply the algorithms to achieve results. This does not require foundational reflection. At the very least, the idea of some kind of *consensus* on the foundations in the English community during this time is quite inappropriate. Guicciardini (1989) has done the most detailed investigation to date of the attitudes in the English community. Presenting a variety of primary evidence he states, 'The problem of foundations was never seriously treated before Berkeley', referring to '[T]he careless approach to foundations of the fluxionists before 1734' (1989: 41).[24]

Lavine (1994) makes a similar point, and thus considers Berkeley's own 1735 summary of the state of play to have been 'pretty fair':

> Some fly to proportions between nothings. Some reject quantities because infinitesimal. Others allow only finite quantities and reject them because inconsiderable. Others place the method of fluxions on a foot with that of *exhaustions*, and admit nothing new therein. Some maintain a clear conception of fluxions. Others hold they can demonstrate about things incomprehensible. Some would prove the algorithm of fluxions by *reductio ad absurdum*; others *a priori*. Some hold the evanescent increments to be real quantities, some to be nothings, some to be limits. As many men, so many minds.... Some insist the conclusions are true, and therefore the principles.... Lastly several... frankly owned the objections to be unanswerable. (Cited in Lavine 1994: 25)

If this is a fair representation, then even if one position or another *was* inconsistent, we would be hard pushed to attach any significance to that position. Further, it is revealing that despite all the confusion and differences of opinion on the foundations, individuals in the English community nevertheless made derivations in the same way, following the same algorithms. This shows vividly how clear the divide is between the foundations and the mathematical practice: even with an inconsistency at the level of justification there was no chance of deriving a contradiction at the *mathematical* level, at least if one stuck to the algorithms. But this is *even if* there is an inconsistency; note that Berkeley, in the above quotation, doesn't mention inconsistency, and many passages in *The Analyst* refer not to inconsistency, but to 'confusion', 'obscurity', and 'incomprehensible metaphysics' (e.g. §XLIX). It might be right to say that there were *some*

---

[24] See Guicciardini (1989) for the details; see also Cajori (1917: 145f.) and Boyer (1949: 222).

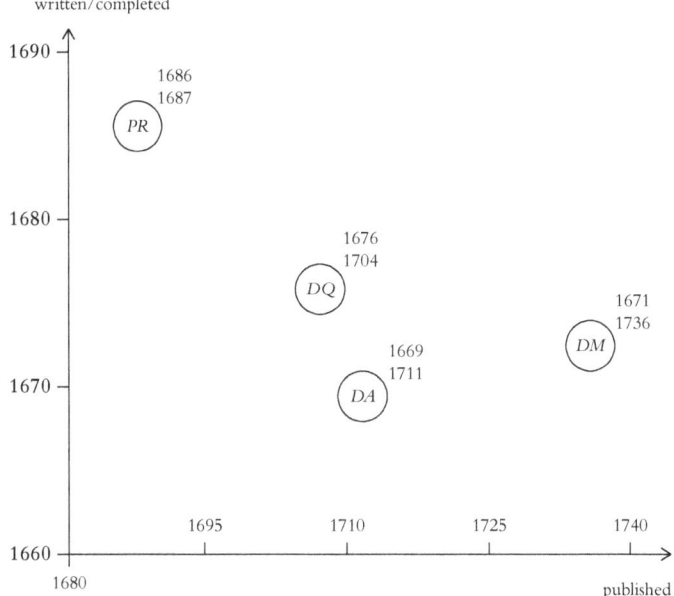

Figure 6.2 The relationship between Newton's major works on the calculus, comparing date written/completed with date published. One sees clearly that the two works published first were written last (*PR* and *DQ*: *Principia* and *De Quadratura*), and the two works published last were written first (*DA* and *DM*: *De Analysi* and *De methodis serierum et fluxionum*). The numbers attending each bubble give first the date written/completed, then the date published

conceptual confusions in the community that amounted to inconsistencies, but then we are really talking about the metaphysical assumptions of isolated individuals and not about 'the calculus' on any possible understanding of that term.[25]

After 1734 there was increased debate on the foundations in an attempt to quash Berkeley's attack. However, it is still very difficult to find anything like a consensus after this time, and very easy to find disagreement, confusion, and metaphysical incoherence. The order of publication of Newton's major mathematical works did not help here (see Figure 6.2). For example, the last publication in Newton's lifetime—*De Analysi* (1711)—presents the

---

[25] Of course Berkeley is pointing out an inconsistency in the sense that different individuals say contrary things, but this is not what is supposed to distinguish the early calculus as particularly interesting and important. This will happen in the 'context of discovery' of just about any new development in science or mathematics.

earliest stage in Newton's development of thought, being written as it was around 1669. Here there is more talk of infinitesimals (cf. Boyer 1949: 191 and 193f.; Guicciardini 1989: 41ff.; Lavine 1994: 16f.). As I said in §6.4.1, before the publication of *De Analysi* Newton did remove *some* talk of 'infinitesimals', but certainly not all such talk. Then shortly after Berkeley's attack, in 1736, Newton's *De methodis serierum et fluxionum* was posthumously published, although this again was written at a very early stage in Newton's development of thought, before he had what he considered a suitable story to tell about the metaphysical foundations of the calculus. Newton's most considered view, in the *Principia*, had been published nearly 50 years before Berkeley's *Analyst*, of course, so compared with *De methodis* it was old news.

With this messy influence from Newton's works there is little wonder that divergence of opinion and confusion was rife. Indeed, after Berkeley's publications in 1734 and 1735 he sat back and let the fluxionists argue amongst themselves. As Cajori (1917) relates,

About twenty articles were written, one of which filled one hundred and thirty-six pages. All articles taken together covered over seven hundred printed pages. They were attempts to ascertain what Newton's ideas of *fluxions* and *moments* were, and whether, Newton meant that a variable can reach its limit or cannot. (Cajori 1917: 149)

This said, there was at least a widespread commitment to the *terminology* of fluxions. As noted in §6.4.1, a commitment to fluxions and a kinematic conception of curves does not determine whether one is also committed to infinitesimals. It remains possible that one wishes to introduce infinitesimal increments of time o, and thus infinitesimal 'moments' taking the form $\dot{x}o$ for a given fluent x. Many fluxionists before 1734 did talk in this way. However there was a fairly widespread and explicit rejection of infinitesimals by the fluxionists in the years after *The Analyst*.

No doubt the publication of Newton's *De methodis* in 1736 was influential. But just as influential was Maclaurin's *Treatise of Fluxions* of 1742. In it Maclaurin explicitly rejected infinitesimals, perhaps influenced by statements in Newton such as the following from *De Quadratura*:

I have sought to demonstrate that in the method of fluxions it is not necessary to introduce into geometry infinitely small figures. (Cited in Boyer 1949: 202)

Kitcher (1973) argues that fluxions remain at the 'algorithmic level', as essentially heuristic and methodological devices, and not as justificational devices. This may have been Newton's own view, but it remains true that many fluxionists post-Berkeley did want to justify the calculus in terms of time and motion. And Maclaurin (1742), in presenting an extended axiomatic approach to the continuity of time and motion, was a clear authority here. After 1742 a great number of textbooks emerged on the calculus, and Maclaurin played a part in all of them. In his careful study of English opinion at this time Guicciardini (1989) writes,

With few exceptions, Maclaurin's view of the calculus dominated the second half of the century: the infinitesimals and the moments of the early fluxionists were definitively abandoned. (p.47)

Since 1742 [Maclaurin] almost all the fluxionists accepted Maclaurin's rejection of infinitesimals. (p.51)

In all these textbooks the reader was introduced in a preface or first chapter to the kinematic meaning of the concepts of the calculus; here Maclaurin was followed as the authority on these foundational aspects. (p.59)

Without infinitesimals how was one to justify the calculus? Here, Newton's method of 'prime and ultimate ratios' did not play a major role.[26] Instead, one finds some rather awkward metaphysical talk about the continuity of time and motion, the concept of velocity at a single 'point' in time, and so on. A typical textbook passage reads as follows:

Let a hollow Cylinder be filled with Water, and let it flow freely out through a Hole at the Bottom of it.... [T]here can be no two moments of Time, succeeding each other so nearly, wherein the velocity of the running Water is precisely the same. And therefore the Velocity that the effluent Water has at any given Point in Time, belongs only to that one particular, indivisible Moment of Time, and no other: And this is accurately the fluxion of the Fluid flowing out at that Moment of Time. Now if precisely at that Moment you begin and continue to pour more Water into the Cylinder, so that the surface of the Water may descend no lower, but keep its Place; then the effluent Water will also retain its Velocity, and continue

---

[26] James Jurin, in his reply to Berkeley on behalf of the fluxionists, did refer Berkeley to Book 1 of the *Principia*, and to a justification in terms of the error being less than any possible given error, and therefore zero. In his reply Berkeley wrote, '[T]o argue that quantities must be equal because they have no assignable difference seems the most injudicious step that could be taken' (see Brook 1973: 187f.). At this point Berkeley's objection is that the justification Newton gives is inadequate, not inconsistent.

to be the Fluxion of the Fluid as before. Now these are the genuine effects and operations of nature itself, and do, in a manner visibly, confirm the truth of what has been said of the Nature of *Fluxions*.' (Emerson 1743: ix–x, cited in Guicciardini 1989: 45)

As Guicciardini puts it, 'There is a circularity in basing the calculus, a mathematical tool devised to study kinematics, on the concepts of time and velocity' (1989: 45). However, here we are clearly moving into the territory of inadequate and incoherent justifications, as opposed to inconsistent ones.

I should also emphasize that very many in the community still didn't care much for foundational talk after Berkeley. Many textbooks appealed to Maclaurin's *Treatise of Fluxions*, as noted, but it isn't even clear that the writers of these textbooks had *read* his book. It has been written that Maclaurin was 'praised by all, read by none' (Grabiner 1997: 394). Again, the real work was considered to be the *use* of the calculus. Speaking about the vast number of textbooks emerging at this time, Guicciardini (1989) writes,

All these treatises, however advanced they may have been, did not introduce the student to the calculus as a theory..., but rather explained to him how to employ in geometry and mechanics a set of rules. (Guicciardini 1989: 58f.)

In a community like this, with many simply following algorithms and not caring about foundations at all, and with those who did care about foundations explicitly rejecting infinitesimals, what sense can we make of the common claim that 'the early calculus was inconsistent', with the contradictory properties of the infinitesimal cited as evidence? I will return to this question in §6.5.

6.4.4 *The French*

In the previous three subsections I have looked at the justificational assumptions made by Newton, Leibniz, and the English mathematical community. Of course I can only provide a flavour of this complex history, but hopefully I have done enough at least to show that any charge of inconsistency must be very carefully interpreted. Turning now to the French community, things are no less complicated, but again various historical facts can be presented which, at the very least, cast a question mark on the suggestion that there is a group—the so-called 'French infinitesimalists'—that made a serious commitment to infinitesimals along with their contradictory properties.

When one thinks of the French community at this time—including those not based in France but nevertheless in close contact with the French—relevant names include Johann Bernoulli, Jakob Bernoulli, l'Hôpital, Malebranche, Varignon, Fontenelle, Saurin, Carré, Montmort, Guisnée, Grandi, Rolle, Ph. de la Hire, Galloys, Gouye, Wolff, Hermann, Euler, and of course Leibniz. Certainly not all of these individuals can be described as 'infinitesimalists': some of them argued explicitly and publicly *against* infinitesimals (including Leibniz as we have already seen). However, many of them articulated a belief in infinitesimals at one time or another. As Mancosu (1989) puts it,

This belief in the existence of infinitesimals was common to all the French infinitesimalists and they shared it with (and probably got it from) Johann (I) Bernoulli. (p.234)

But of course if we take this at face value there is a clear sense in which it is paradoxical. Are we really to accept that many of the prominent mathematicians and intellectuals just mentioned believed what is obviously impossible: that infinitesimals exist, along with their contradictory properties?

We can begin to make some sense of this puzzle by taking a look at a few of the individuals just named. First of all, as I mentioned, several individuals explicitly rejected infinitesimals, and so one certainly cannot speak of a consensus in the French community. As Mancosu (1989) writes, in the first decade of the eighteenth century 'a bitter debate raged for about 6 years' within the Academy of Sciences in Paris. Michel Rolle was one of the frontrunners, but he was supported by Ph. de la Hire, Galloys, Gouye, and a few others. In a moment I'll turn to the reaction of some of the 'core' infinitesimalists at this time. But first, consider a second sub-group within the community: those who deferred to Leibniz. As noted in §6.4.2, Leibniz had responded to criticisms courtesy of Nieuwentijt in 1694 and 1695. When Nieuwentijt responded to Leibniz, and it was clear Leibniz was not moved to respond again, Jacob Hermann (a disciple of Jakob Bernoulli) stepped in. Part of the defence was to state explicitly Leibniz's own generalization of equality as a basic assumption: 'Whatever differs by a difference smaller than any given quantity is equal' (cited in Nagel 2008; 206). And

Hermann was not the only one to take an essentially Leibnizian position. Later, in 1730, the German philosopher—and long-time correspondent of Johann Bernoulli—Christian Wolff wrote, 'Infinitely small quantities are impossible and thus the infinite or the infinitely small quantities of the mathematicians are not real quantities but only imaginary ones' (cited in Nagel 2008: 211).

So we have those in the community who rejected infinitesimals, and those who perhaps accepted the *terminology* of infinitesimals but took a Leibnizian stance vis-à-vis foundations. A third subsection of the community worth mentioning are those who just didn't spend any time worrying about justification, or at least the justification of the moves made *within* the algorithms. On the Continent, as in Britain, straightforward *use* of algorithms without concern for foundations was rife. Many in the community, even those who were especially prolific, never wrote anything public on the foundations, and hardly ever wrote anything in personal letters on the foundations. This is certainly true of the Bernoulli brothers, even if they are usually identified as 'core' infinitesimalists. As Nagel (2008) puts it,

> Soon both Bernoulli brothers could handle the new method. But of course they were both primarily interested in solving a large number of the most difficult problems in mathematics and physics, and not in meditating on the basic notions of the calculus and on their rigorous foundation. (Nagel 2008: 199)

Euler came a generation later, but he fits in here too. Boyer (1949) also adds a few more names to this list, writing,

> [T]he greater the success achieved by the differential calculus, the less constrained Euler felt to justify his procedures. His views on the bases of the subject were elementary in the extreme, resembling somewhat those of Wallis, Taylor, John [Johann] Bernoulli, and Fontenelle. (Boyer 1949: 243f.)[27]

Another 'French infinitesimalist' who really had no interest in the foundational metaphysics was the Marquis de l'Hôpital. His major publication was the 1696 textbook *Analyse des infiniment petits*, or 'Analysis of the Infinitely Small'. The title obviously suggests that l'Hôpital believed in the existence

---

[27] What Nagel (2008) writes of Johann Bernoulli can perhaps be said of each of these individuals: 'Utility and success in calculation are decisive for him. And as long as he gets convincing results he does not care about conceptual deficiencies' (2008: 213). In some cases one might even say 'doesn't *notice* conceptual deficiencies'. See Boyer (1949) for a discussion of Wallis, Taylor, and Fontenelle.

of infinitesimals, but in fact things are not so clear. L'Hôpital does not discuss the foundations of the calculus in his book at all (see Boyer 1949: 238). In fact he intentionally avoids metaphysical talk. There is a curious story behind this. L'Hôpital really got his ideas from Johann Bernoulli: by 1691 Bernoulli had entered into a contract with the Marquis whereby it was agreed that, for a handsome sum, Bernoulli would communicate all his ideas directly and exclusively to l'Hôpital.[28] When he did so, Bernoulli set down a quite paradoxical claim:

1. A quantity, which is diminished or increased by an infinitely small quantity, is neither diminished nor increased.[29]

But when it came to his textbook, L'Hôpital changed this to the following:

Grant that two quantities, whose difference is an infinitely small quantity, *may be taken* (or used) indifferently for each other: or (which is the same thing) that a quantity, which is increased or decreased only by an infinitely small quantity, *may be considered* as remaining the same. (Mancosu 1989: 227, emphases added)

As Nagel (2008) has written, L'Hôpital makes a small but significant change here. In Nagel's words,

[T]he difference could be interpreted as if Bernoulli treated the entities to which the notions of the calculus refer as real objects, whereas L'Hôpital speaks of them hypothetically, as if he were using the notions of the calculus as a mere manner of speaking. (Nagel 2008: 199f.)

It is curious to refer to L'Hôpital as an infinitesimalist on the grounds that he 'believed in infinitesimals', when in fact he had no interest in, and goes out of his way to avoid, foundational issues.

What then of Johann Bernoulli? We saw in §6.4.2 that he argued with Leibniz in 1689 that infinitesimals exist. One option for individuals who declared 'infinitesimals exist' is that they actually had a conception of infinitesimals which somehow avoids the contradictory properties—in other words although it might be right to label them 'infinitesimalists', this doesn't mean they had inconsistent commitments. For example, some members of the community took differentials to be variables, not constant,

---

[28] For more on this relationship see Bos (1980: 52 and 73).
[29] '1. Eine Größe, die vermindert oder vermehrt wird um eine unendlich kleinere Größe, wird weder vermindert noch vermehrt' (Bernoulli 1924 [1691/2]).

infinitesimal quantities (as we will see very shortly). This is *not* true of Johann Bernoulli: he stated explicitly that infinitesimals are constant quantities.[30] Then to top it off we have the paradoxical statement given above: quantities which *are* increased by an infinitely small quantity are *not* increased. And this was not just a throwaway remark, but was set down as a Postulate right at the beginning of an essay on the 'Differential Calculus' which he wrote in 1691.

Here we have what is surely one of the most radical positions in the history of science and mathematics: Bernoulli seems quite clearly to embrace an obvious contradiction. In this he was really quite on his own, as we might expect. Often he is grouped together with his brother Jakob. For example Edwards (1979) writes,

[W]hereas Leibniz himself was somewhat circumspect regarding the actual existence of infinitesimals, this appropriate caution was generally not shared by his immediate followers (such as the Bernoulli brothers), who uncritically accepted infinitesimals as genuine mathematical entities. (Edwards 1979: 265)

However, although Jakob may have believed in infinitesimals *in a sense*, he stated on at least one occasion that a quantity smaller than any given magnitude is zero (Boyer 1949: 240). This is not what his brother would have said: for Johann infinitesimals just were quantities smaller than any given quantity. At other times Jakob did appear to commit to infinitesimals (Boyer 1949: 240; Mancosu 1989: 238), but this may be because he followed others in the French community who conceived of infinitesimals as *variable* quantities.

When it comes to 'infinitesimals as variable quantities' the most important figure is Varignon. He is so important, since during the 'bitter debate' at the beginning of the eighteenth century Varignon stepped up for the infinitesimalists, as a representative, to defend their position (see Mancosu 1989). Michel Rolle provided the familiar objections:

A quantity + or − its differential is made equal to the very same quantity, which is the same thing as saying that the part is equal to the whole.... Sometimes the differentials are used as nonzero quantities and sometimes as absolute zeros. (Mancosu 1989: 230)

---

[30] See Nagel (2008: 213). There is evidence that Johann Bernoulli's conception of the infinitesimal did not change between 1691 and 1730.

However, instead of defending the inconsistent properties of infinitesimals, Varignon rejected the suggestion that infinitesimals were constant quantities at all. He wrote to Leibniz in 1701,

The Abbé Galloys, who is really behind the whole thing, is spreading the report here [in Paris] that you have explained that you mean by the 'differential' or the 'infinitely small' a very small, but nevertheless constant and definite, quantity.... I, on the other hand, have called a thing infinitely small, or the differential of a quantity, if that quantity is inexhaustible in comparison with the thing. (Cited in Boyer 1968: 475)

As Boyer writes at this point,

The view that Varignon expressed here is far from clear, but at least he recognized that a differential is a variable rather than a constant. (Boyer 1968: 475)

At the very least, Varignon goes directly against Johann Bernoulli when he rejects the suggestion that 'the "differential" or the "infinitely small" [is] a very small, but nevertheless constant and definite, quantity'.

The view of differentials as variable quantities bears more than a passing resemblance to Newton's 'flowing moments'. Surprisingly enough, in an effort to prove that infinitesimals exist, Varignon turned directly to Newton's fluxions and an explanation in terms of kinematics. As Mancosu puts it, he 'appealed to Newton's *Principia* as the source for a rigorous foundation of the calculus' (1989: 231). At one stage he even gives an explanation reminiscent of Newton's ultimate ratios, citing infinitesimals as 'infinitely changing until zero'. He goes on,

[T]hey (the differentials) are always real and subdivisible to infinity, until in the end they have completely ceased to exist; and that is the only point at which they change into absolutely nothing. (Mancosu 1989: 231)

Within this web of 'incomprehensible metaphysics', as Berkeley might have put it, one thing is clear: Varignon is concerned to show that the inconsistency claim is not relevant to his position. So if Varignon is representative of the infinitesimalists, this community is *also* not guilty of committing—at least in a doxastic sense—to the inconsistent justification put forward by Berkeley. And he does seem to be representative: that the majority of the community took differentials to be variable rather than fixed quantities is stressed by Bos (1974: 17). In 1710 Leibniz wrote, '[T]he quantity dx itself is not always constant, but usually increases or decreases continually' (Bos 1974: 17). In a discussion of Leibniz, Bassler (2008) writes,

> We should not think of differentials such as dx or dy as quantities but rather as variables, just as x and y are thought of as variables. It is particular *instances* of these variables which are themselves quantities. (Bassler 2008: 142, fn.11)[31]

Breger (2008: 195f.) has also emphasized this conception of the infinitesimal as a variable. He writes, 'If one misunderstands the differentials as "genuine mathematical entities" and as "fixed, but infinitely small," the infinitesimal calculus naturally appears to have "inconsistencies".'

If the differential is a variable, how does the calculus make sense? It has been claimed that it doesn't. Mancosu (1989) writes, 'dx functioned as a numerical constant, and, interpreting it as a process, Varignon's approach created an asymmetry, an incongruity, between the formalism and its referents' (1989: 235). Similarly, Colyvan (2009) writes,

> [I]f an infinitesimal, δ, is a changing quantity, it cannot appear in equations such as: $a = a + \delta$ where a is a constant. Why? Well, the term on the right is changing (since δ is changing) so cannot equal anything fixed, such as a constant a. (§2)

The first point to make here for my purposes is that, if this is right, this means there isn't an inconsistency at the level of justification. Instead the attempted justification doesn't explain the moves made within the algorithm after all; in other words it is inadequate. Perhaps one could also make a case that it is incoherent. But the inconsistency due to the infinitesimal being both zero and non-zero disappears if the infinitesimal is not conceived as a constant quantity.

But in addition we might wonder whether Colyvan's point is fair. We saw above in §6.3 that if we are allowed to form the equation $x + o = x$, where $o \neq 0$, then we can derive $2 = 1$. So regardless of whether o is a constant or a variable, the formulation of '$x + o = x$' is incoherent. As I said above, this just isn't how people reasoned in the days of the early calculus. Rather, they reasoned using algorithms, following sets of ordered rules. '$x + o = x$' just never got written down during these calculations, and neither did

---

[31] There is some disagreement in the literature here, possibly because there is no question of how we 'should' think of the differentials, since some in the community thought of them as variables and some thought of them as constant quantities.

anybody believe it to be true (except perhaps Johann Bernoulli). In addition, by 'o is a variable' people did *not* mean that it can be taken to vary within a single equation (this would be an absurdity, as Colyvan claims), but rather that it can be taken to vary *within a whole procedure*. In other words, one imagines carrying out the whole procedure again and again with o getting smaller each time (e.g. because it is being halved again and again, cf. Table 6.1), and we consider what happens to the final result. In any individual instantiation of the procedure, the infinitesimal (or the 'differential') is a constant.[32]

What of the 'final victory of the infinitesimal calculus' over its critics in 1707 (Mancosu 1989: 243)? What must be noted here is that by this time the battle had turned from justificatory issues to issues of application. Rolle now claimed that the new calculus led to mistakes and he brought forth example after example in an attempt to show this. Sadly for him, in every case it was shown that the new calculus *did* give the right answer and that Rolle had made a mistake in applying it (Mancosu 1989: 232–40). With this strategy in place there is little wonder that pressure mounted up against Rolle. In the face of the common threat, those who had different justificatory stories to tell naturally joined forces to defend the calculus as a *method*. Berkeley was extremely shrewd when it came to this point, emphasizing again and again that in his view 'Analysts arrive at truth'. He wrote,

I have no controversy about your conclusions, but only about your logic and method. How you demonstrate? What objects you are conversant with, and whether you conceive them clearly? What principles you proceed upon; how sound they may be; and how you apply them? It must be remembered that I am not concerned about the truth of your theorems, but only about the way of coming at them. (*Analyst*, §XX)

It is a shame that Rolle didn't do the same: if he had perhaps the French community would have been pushed to clarify things a little further. As it happened, the 'victory' of 1707 was a victory of application, a victory of terminology and method, but not a victory of justification. The wide variety of different justificatory stories remained after 1707. For the specific

---

[32] Kitcher (1973) does a nice job of bringing out how o features as a variable in the context of Newton's fluxions (which also helps us to make sense of Varignon's reference to Newton). See especially p.47: 'The proof then proceeded by treating the "Augment" as a variable dependent upon time...'

case of Leibniz this is brought out vividly in various letters. As Leibniz wrote in 1716 in a letter to a French mathematician named Pierre Dangicourt,

> When they [our friends] were disputing in France with the Abbé Gallois, father Gouye and others, I told them that I did not believe that there were actually infinite or actually infinitesimal quantities; the latter, like the imaginary roots of algebra ($\sqrt{-1}$), were only fictions, which however could be used for the sake of brevity or in order to speak universally. ... But as the Marquis de l'Hôpital thought that by this I should betray the cause, they asked me to say nothing about it, except what I had already said in the Leipzig *Acta*, and it was easy for me to comply with their request. (In Jesseph 2008: 230f.)

In other words, since Leibniz was much more concerned that the calculus be accepted as a method for deriving results, he was happy to keep his foundational thoughts to himself at this time.

6.5 'As if' Inconsistency

In previous case studies I assumed that the two interesting and important ways in which inconsistency can manifest itself in science are (i) when scientists (knowingly) reason with inconsistent propositions, but nevertheless avoid ECQ and are willing to trust at least some of their inferences, and (ii) when scientists are (unknowingly) making a doxastic commitment to inconsistent propositions, and have to work out how to revise those beliefs when the inconsistency comes to light. In the case study currently under discussion I have covered both of these cases. In the early calculus, scientists couldn't possibly derive a contradiction if they were making derivations by applying the relevant algorithms. Indeed, they were not reasoning with inconsistent propositions at all in such a case, but rather following a procedure. And I tried to present enough of the relevant history and relevant historical scholarship to show that Newton and Leibniz did not have inconsistent beliefs, and neither did the majority of their followers. Even those who claimed to 'believe in the existence of infinitesimals' often interpreted 'infinitesimal' in such a way that the obvious contradiction was avoided (even if this meant that their metaphysical beliefs failed to justify their practice at the algorithmic level).

This in itself is enough to cast a question mark on much of the philosophy of science literature which helps itself to the notion that 'the early calculus was inconsistent'. This case is too idiosyncratic, and too complex, for this to be a sensible thing to say without careful clarification. But if we dig a little deeper there may still be a sense in which there is an interesting and important inconsistency here. In short, we might wonder whether we can describe the early calculus as a case of 'as if' inconsistency, where the communities in question made an 'as if' commitment to inconsistent propositions. This is trivial in the case of Johann Bernoulli: he reasoned *as if* he believed in self-contradictory infinitesimals because he *did* believe in self-contradictory infinitesimals. But it's also true of Leibniz: regardless of the careful justifications in his DQA and elsewhere, he still reasoned *as if* he believed in infinitesimals, and in fact he said explicitly that he set out to do just this. Newton, we saw, had a similar story to tell about the role of infinitesimals, and one can certainly find examples in Newton's works (e.g. Kitcher 1973: 43–4) where the moves within a derivation 'are defended only through the invocation of infinitesimals' (Kitcher 1973: 43–4). And clearly this is also the case for many of those working in the English and French communities during this period: even if they had some subtle story to tell about infinitesimals, or perhaps no justificatory story at all, they still often reasoned *as if* they believed in infinitesimals, with their contradictory properties. Didn't they?

Well perhaps not. We've seen that mathematicians followed algorithms, and then (sometimes) attempted to provide justifications for the moves made within those algorithms. Based on this picture one might argue as follows: It might look as if they were reasoning with inconsistent propositions, but this is just an illusion: they were not. And—if we're going to learn anything about how science works—it's not interesting to consider the way it *looks* like people were reasoning; it's interesting to consider the way they *really were* reasoning. As Jesseph (1993) writes,

> His [Berkeley's] lemma that inconsistent suppositions are not admissible in a mathematical demonstration is clearly correct, but Berkeley has shown only that there is the *appearance* of inconsistency in the handling of infinitesimals and evanescent increments. (Jesseph 1993: 227, original emphasis)

Prima facie, it's difficult to see what might be interesting or important about the fact that somebody *appears* to reason in an inconsistent way if in fact they are not reasoning in that way, and don't have inconsistent beliefs. This would be rather like the example I mentioned in Chapter 3 of somebody stating 'I can't believe he didn't come!' when they obviously *do* believe he didn't come. The only inconsistency comes from an inappropriate, overly literal reading of 'I can't believe he didn't come'. Similarly, we might say, in the case of infinitesimals.

However, this argument depends on a certain take on what we mean by 'reasoning' which might be rejected. Take Leibniz's fictionalism: Leibniz declared that infinitesimals were well founded fictions, so that it was fine to reason in terms of them, whilst ultimately preferring a justification which speaks only of finite quantities. But this still seems to mean that there is a very real sense in which Leibniz *was* reasoning in terms of infinitesimals, just like Johann Bernoulli was. Where Bernoulli thought to himself in the process of a proof 'we now increase this quantity by an infinitesimal...' so too did Leibniz. It would seem that they were thinking in exactly the same way as they reasoned. The only difference is that, if suddenly asked to switch their thoughts from procedure and derivation to metaphysics and justification, Bernoulli would carry on thinking about infinitesimals and Leibniz would not. But if asked to discuss a derivation Leibniz might suspend his belief that infinitesimals are fictions and converse with Bernoulli just as if he, like Bernoulli, believed that they exist. Similarly we could have a perfectly sensible conversation with somebody who (bizarrely) believes the film *Inception* is relating true events, about whether or not Cobb is still in a dream in the final scene. We can do this even though what we really believe is that the film does *not* relate true events, and the filmmakers have left the question open for the viewer to decide.[33]

It would still be the case that Leibniz does not have an inconsistent belief set. It would be closer to the truth to say that he is *pretending* he does for the sake of carrying out derivations, because it is heuristically useful and simplifies things. But actually even this does not seem right. Suppose we pretend

---

[33] Some have claimed that to believe something *just is* to act as if it is true; cf. criticisms of van Fraassen's distinction between belief and acceptance by Horwich (1991) and others. I side with the critics of these criticisms, maintaining that in order for this to work one must build in other, objectionable assumptions, e.g. concerning behaviourism (cf. Maher 1990; Ketland 2005). I take it to be obvious that just because one acts *as if* the 'increment' o is both non-zero and then also zero, one doesn't *necessarily* believe there is this thing, the infinitesimal, which both is zero and isn't zero. Newton and Leibniz bring the distinction to life.

(it is true that) infinitesimals exist along with their contradictory properties. If, in that pretend scenario, we also believe that classical logic is truth preserving, then by ECQ we must also pretend to believe that everything is true, including '2 = 1'. To avoid this we will have to also pretend that some basic rules of logic do not hold. Further, in the pretend world we've created we must reason as follows at the beginning of a derivation:

- The infinitesimal is both zero and a non-zero quantity.
- Therefore (by &-elimination) it is a non-zero quantity.

And at the end of a derivation we must reason as follows:

- The infinitesimal is both zero and a non-zero quantity.
- Therefore (by &-elimination) it is zero.

There has to be some reason why we can only do the former &-elimination move at the *beginning* of a derivation, and the latter &-elimination move at the end. Indeed, we need to create this pretend world in just such a way that one reasons by the algorithms and not otherwise. As we've already seen, those who used assumptions about infinitesimals to depart from the algorithms (e.g. Michel Rolle) were quickly shot down by the community.

This fictional world with self-contradictory infinitesimals, complex inferential restrictions, etc., does not make sense of the fact that Leibniz and others thought in terms of infinitesimals as a helpful way to simplify the metaphysical and logical picture during a derivation. And there is an alternative way to think about Leibniz's 'as if' commitment which *does* make sense of that fact. At the beginning of a derivation Leibniz does not need to pretend there are infinitesimals which are both zero and non-zero and then infer from this (by &-elimination) that they are non-zero. He can just go ahead and pretend they are non-zero. Later in the derivation he can pretend that they are zero. This shift in what one is pretending is harmless: one can pretend that infinitesimals are non-zero at one point and then pretend they are zero at another point without thereby being committed to pretending that they are both zero and non-zero at any time. Thus in this case there is no motivation for a paraconsistent 'logic of pretence', since there is never any inconsistent 'as if' commitment.

Here Leibniz and Bernoulli come apart, even at the level of the story they tell themselves as they carry out a derivation. Bernoulli, in believing two inconsistent conjuncts at slightly different times, is committed to believing the full conjunction at a *single* time. This is just how belief works: one can opt in and out of *pretending* to believe something, but not so with *really believing* something. Beliefs are something much more difficult to shift: not only can one not justify changing beliefs so quickly, on most theories of belief it is psychologically *impossible* to change one's beliefs just like that. Obviously it is psychologically possible to change what you're *pretending* at will. But suppose somebody asked 'what *justifies* changing what you're pretending?' At this stage, the mathematician in question can launch into his or her preferred justificational story: Leibniz can talk about his fictionalism and the generalization of equality in his DQA; Newton can talk about prime and ultimate ratios and 'limits'; fluxionists can talk about time, velocity, and continuity, and so on.[34] In fact, one might also put in this category some of those individuals without any proper justification. In answer to the question 'what justifies changing what you're pretending?' they would answer 'well, I just know it works out if I do that'.[35]

So we have an obvious alternative to the story where Leibniz et al. are pretending infinitesimals exist, along with their contradictory properties. I don't take this to be absolutely definitive: this talk of 'the story one tells oneself' as one carries out a derivation is a delicate question in the psychology of mathematics. Perhaps both of the above mentioned reconstructions are permissible, although as I have indicated I believe the latter to be more veridical. At the very least we see that the simple story we are met with so often in the literature, of the early calculus as a set of inconsistent propositions plus a logic, is plain wrong. Brown and Priest (2004) are typical of a subsection of philosophy of science that assumes the early calculus can be reconstructed by making use of a paraconsistent logic. To motivate the application of a particular paraconsistent logic they dub 'chunk and permeate' they write,

---

[34] Maclaurin (1742) might also have respected fictionalism (even if he didn't practise it): 'I have always represented Fluxions by all Orders of finite Quantities, the Supposition of an infinitely little Magnitude being too bold a *Postulatum* for such a Science as Geometry. But because the Method of infinitesimals is much in use, and is valued for its conciseness, I thought it was requisite to account explicitly for the truth, and perfect accuracy of the conclusions that are derived from it' (cited in Guicciardini 1989: 49).

[35] The question may arise of whether they really *are* pretending when they don't have an alternative story to tell. They might well reply that of course they are, otherwise they would believe something contradictory. On many theories of belief it is not just bizarre to believe something contradictory, it is impossible to do so.

> The infinitesimal calculus of Leibniz and Newton is well known to have operated on an inconsistent basis. In particular, at different points in the calculation of a derivative infinitesimals had to be assumed to be both zero and non-zero. How was the trick turned? The fact that arbitrary conclusions were not drawn, even though reasoning was carried out on the basis of inconsistencies, means that the inference procedure involved must have been paraconsistent. (Brown and Priest 2004: 379)

Clearly this blurs the important distinctions between the algorithms of the calculus, the story one tells oneself whilst making a derivation, and the attempted justifications of the moves made within the algorithms. First, to say of Leibniz that he 'assumed' that infinitesimals had to be both zero and non-zero—indeed that he *had* to assume that—is to completely ignore his fictionalism (similarly with Newton and others). Second, to claim that 'reasoning was carried out on the basis of inconsistencies' is to ignore the fact that mathematicians made derivations by following algorithms. Third, even at the level of 'as if' commitment I have just shown that a paraconsistent inference procedure was *not* necessary: instead one can reconstruct the process as a change in what one is pretending (of course, this depends on *not* ignoring the role of fictionalism). Fourth, even if we allow that people *were* inconsistently committed at the 'as if' level, a paraconsistent reconstruction is not motivated. This is because it was the algorithms that dictated what mathematicians should do during a calculation, not the story mathematicians told themselves as they carried out such calculations. The fictional story was built to match the algorithm, not the other way around: in other words it did not have influence over inferences. It really was *just* a story one told oneself as a helpful heuristic device (apart from the case of Johann Bernoulli, perhaps). One does not need a paraconsistent logic to explain why mathematicians did not derive arbitrary conclusions: they didn't derive arbitrary conclusions because their inferences were dictated by canonical algorithms in the works of Newton, Leibniz, and the textbooks which followed.

Finally, it is important to emphasize that many mathematicians didn't even *pretend* they believed in infinitesimals to help them work through proofs. Consider (i) those who really did just blindly follow the rules of the algorithm, without thinking in terms of infinitesimals at all, (ii) the fluxionists who rejected infinitesimals and reasoned by making use of the concepts of time, velocity, and continuity, and (iii) those who worked in terms of infinitesimals *as variables*, such as Varignon. In the latter two cases, whether the justification works or not is not the point. The point is that, for these individuals, there is no distinction made between the justification of the algorithm and the story they tell themselves as they work through the algorithm. So if their justification isn't inconsistent, neither is this story.

6.6 Conclusion

Clearly the early calculus is a complex case. Any story which suggests that it can be represented simply as 'propositions plus logic' does not do any justice to this complexity, and arguably tells us nothing about how real science/mathematics works. Of course, this doesn't mean that there can't be an interesting inconsistency which is *somehow* associated with the early calculus. Given my method of theory eliminativism I am accepting the challenge of identifying the propositions which are supposed to be inconsistent, and then considering all of the ways in which this inconsistency *could* be interesting or important given the true complexity of the history.

This has led me to consider three aspects of the early calculus: (i) the algorithms used to make derivations, (ii) the attempted justifications of those algorithms, and (iii) the story one tells oneself as one carries out a derivation (including the 'as if' commitments of some in the community). First (§6.3) I argued that the algorithms are not 'inconsistent' (in any sensibly adjusted sense of 'inconsistent'). Thereafter (§6.4) I argued that the justifications offered by various individuals in the community were usually inadequate, incoherent, or just non-existent instead of being inconsistent. The best possibility for a significant and widespread inconsistency is in the 'as if' commitments of those working with the algorithms of the calculus (§6.5). But even here things are not so clear. Many in the community did not invoke infinitesimals even at the 'as if' level (e.g. many fluxionists following Maclaurin). Those that clearly did (e.g. Leibniz) did not have inconsistent commitments here if we think of them as changing what they are pretending half way through a derivation. This way of thinking may sound rather odd, but, as I argued in the previous section, the alternative in terms of inconsistent commitments is odder still. Changing what one is pretending

THE EARLY CALCULUS 189

for the sake of convenience is really quite harmless, especially if one has an underlying justification for it, as Leibniz did.

Johann Bernoulli remains as somebody who did commit to infinitesimals in a serious way (although he hardly ever discussed these beliefs, caring more for deriving results). But just to find one individual is certainly interesting. It seems to show that the common fear of inconsistent assumptions is unfounded: Bernoulli proceeded to use the calculus to stunning effect, and was undoubtedly more proficient with it than many who steered clear of his peculiar justificatory beliefs. So isn't it the case that we should learn to embrace inconsistency more readily, that we shouldn't fear the 'logical explosion' classical logic tells us will ensue?

The problem with this story is that there is a clear sense in which even Bernoulli was *not* reasoning with inconsistent propositions: instead, he was following the algorithms. He never veered from these algorithms, despite his apparent belief in infinitesimals as 'real things'. It is possible to explain this by painting Bernoulli as somebody who was *also* committed to rejecting other basic mathematical assumptions, changing the underlying logic, and so on (recall previous section). There may be some value in this picture of things, but again we don't seem to have any historical relevance here. Bernoulli didn't reflect on foundational questions at all, so he hardly thought about things like changing the underlying logic. And the suggestion that he did this subconsciously or implicitly is unpalatable too: the changes required are substantial, not something one could do without noticing!

Some philosophers may find this constant reference back to the 'real history' tiresome. Consider the following passage from the Brown and Priest (2004) article I discussed in the previous section:

First, a word of caution. This is not an historical paper. We do not claim that Leibniz, Newton, and the other mathematicians who worked with the infinitesimal calculus explicitly presented their work in this ['chunk and permeate'] way. There is therefore no point examining texts to determine whether or not they did. The strategy we give is, rather, that of a rational reconstruction. What we do claim is that this strategy captures the practice that they engaged in. (Brown and Priest 2004: 381)

However, I find the terms 'rational reconstruction' and 'captures' utterly mysterious here. Is the claim that mathematicians reasoned in a 'chunk and permeate' way not explicitly, but *im*plicity? Surely not, since then there *would* be a point to examining the texts. (In addition, if my own interpretation of the texts is anything like right, they would *not* show an implicit commitment to a paraconsistent logic.) Is the claim that it is a way individuals *could* have reasoned, even though nobody did (a normative, not a descriptive claim)? This doesn't seem to make sense either, since then how would it 'capture the practice they engaged in'? In the face of such mystery, and absent any further explication, I find no other option than to be sceptical that Brown and Priest have a story to tell here about how exactly their 'rational reconstruction' is informative vis-à-vis how science works, or how it tells us something sensible and realistic about how science *could* work.

However, Brown and Priest are simply following a theme in philosophy of science which is completely entrenched. As I mentioned at the beginning of this chapter, Lakatos, Feyerabend, Shapere, Priest and Routley, and da Costa and French all put 'the early calculus' on lists of 'inconsistent scientific theories'. Da Costa and French provide a 'paraconsistent logic of acceptance' for handling such theories. In two recent papers Colyvan writes as follows,

Take the case of the early calculus. This theory was straightforwardly inconsistent: infinitesimals were taken to be non-zero (by stipulation and by the fact that one needed to divide by them) and yet at other times they were taken to be equal to zero. Moreover, within one proof (such as a standard proof from first principles of the derivative of a polynomial, for example) both these contradictory properties were invoked. Newton and Leibniz also seemed to have contradictory interpretations of the infinitesimals. (2008: 118)

A great deal of one of the most important periods in the history of science—the late 17th century to the mid 19th century—relied heavily on inconsistent mathematics. During this period, most scientists were working with an inconsistent mathematical theory [the early calculus] and this theory was used almost everywhere. (2009: §2)

Colyvan is typical here. The early calculus is interpreted as 'a set of propositions plus inference rules'. The inconsistent propositions are said to be 'used' in derivations. And Newton and Leibniz are even said to have 'contradictory interpretations of the infinitesimal'. As we have seen, all of this is extremely contentious, if not plain false. It is striking, to say the least, that there is such a disconnect not just between the real history and philosophy of science, but between *modern historical scholarship* and philosophy of science.

The general objection here is, of course, the basic point that philosophies of science often fail because the full complexity of the science and/or the history isn't taken into account. With this in mind, *can* we hope to say something quite general about 'inconsistencies in science'? Well, perhaps we can, since one might argue that we haven't *seen* many genuine examples of inconsistency in science yet. The Pauli inconsistency in the later Bohr theory and (I2) in Newtonian cosmology are the only cases we might describe as 'clear cut'. About these sorts of cases perhaps we can hope to say something quite general. But if we turn to the list of examples of 'inconsistent science' found in the literature, the possibility of saying something substantial and general looks slim. At the very least, in each of the cases we've seen so far the inconsistency manifests itself in a different way, and the 'early calculus' is perhaps the most idiosyncratic case yet. The reader might think this should have been obvious from the start, simply because here the subject matter is mathematics, and not science. However, this certainly isn't the impression one would get from reading the philosophy of science literature: the 'early calculus' is mentioned as a classic example of 'inconsistent science' nearly as much as Bohr's theory of the atom. Philosophers often group it with other cases, and then try to say something general about the rationality of scientific commitment to inconsistency, how we do, or should, reason with such theories, or how such inconsistencies might be overcome. As I have tried to show in this chapter, we really can't afford to ignore the specifics of the relevant history.

With these ideas in mind I turn in the next chapter to consider more briefly four further examples of apparent inconsistency in science. In the light of the previous four case studies the primary questions will now be: *Is* there really a substantial inconsistency when we consider the relevant science and history in all its complexity? And, if there is: When we compare this case with other cases of inconsistency in science, is there something substantial and general we can say about how science works in such circumstances? Or does it depend on the specifics of the case at hand?[36]

---

[36] A final comment is in order before this chapter closes. At the eleventh hour prior to publication I was urged to consider that a more charitable interpretation of Johann Bernoulli might show *him too* not really to be committed to inconsistent claims. Mikhail Katz (personal communication) urges that Bernoulli relied on crucial clarifications within Leibniz's work (of the sort discussed in §6.4.2) and that his apparently inconsistent commitments should be interpreted in the light of these clarifications. The interested reader is directed to Katz, Schaps, and Shnider (forthcoming) and Katz and Sherry (forthcoming 2013).

7

Further Examples

7.1 Introduction

Several other cases are sometimes labelled 'inconsistent scientific theories', or at least 'examples of inconsistency in science'. Here I will consider four of the most prominent examples: Aristotle's theory of motion (§7.2), Olbers' paradox (§7.3), the theory of classical electrons (§7.4), and Kirchhoff's theory of diffraction (§7.5). This will bring the number of considered cases to eight, and will mean that the thesis discusses a variety of examples spanning a range of historical periods and disciplines. In addition these further case studies stand as tests of the hypotheses which have been formed in the course of the previous four chapters. In particular I will ask, is there really a good reason to refer to these cases as 'inconsistent scientific theories', or even as 'inconsistencies in science'? Or is there a significant disconnect between the philosophy of science and the science itself, as we have seen in some of the previous examples? And if there are genuine inconsistencies here, are the cases especially idiosyncratic, such that substantial, general claims about 'inconsistency in science' are unlikely to be possible?

7.2 Aristotle's Theory of Motion

It is widely written that Galileo showed Aristotle's theory of motion to be internally inconsistent, as giving rise to a contradiction. For example, Priest and Routley (1983) cite Bohr's (1913a) theory of the atom, the Aristotelian theory of motion and the (early) infinitesimal calculus as examples of 'inconsistent theories in science and the history of science' (1983: 188). They intend such examples to provide evidence for the importance of paraconsistent logic within real science. Citing these examples they write,

> Thus, there are genuinely inconsistent theories in the history of science. Moreover, even if our claims specifically about Bohr's and Newton's theories were not correct, the friends of consistency could not claim *a priori* that *prima facie* inconsistent theories always had *ad hoc* consistentizations. That would, after all, beg the question against us. What would be required, rather, would be a detailed historical analysis of many and varied cases of such inconsistent theories. And such an analysis... would, we conjecture, issue in our favour. (Priest and Routley 1983: 188)

Although the applicability of paraconsistent logics has not been my focus, the current investigation does count as something like the 'detailed historical analysis' Priest and Routley have in mind. But so far it hasn't issued in their favour: we have already seen that many of the alleged internal inconsistencies in science are not significant inconsistencies in the way usually assumed. If the inconsistency of Aristotle's theory of motion is similarly insignificant, Priest and Routley's argument for the use of paraconsistent logics in these cases will be dramatically weakened.

A lot of work has already been done on Galileo's criticism of Aristotle's theory of motion. As a thought experiment, the argument runs as follows. Aristotle says that the speed at which a freely falling body falls is proportional to its weight. Now imagine that two bodies of different weights are strapped together, body H being the heavy body and body L being the light body. Since L is naturally inclined to fall at a slower speed than H, when the two bodies are strapped together it will hold H back and the two will fall at an intermediate speed, a speed somewhere in-between the natural speed of L and the natural speed of H. But also, because we now have a total body L+H which is obviously heavier than H, the strapped body should fall at a speed faster than H. So we have a contradiction: L+H should fall both slower and faster than H would fall on its own. Therefore, it is claimed, we must reject our original supposition that a freely falling body falls with a speed proportional to its weight.

The thought experiment can be reconstructed as a more formal argument (cf. Gendler 1998: 404; Schrenk 2004). The theory is presented as follows:

(i) Natural speed is directly proportional to weight.
(ii) Weight is additive.
(iii) Natural speed is mediative.

From (ii) and (iii) we get the negation of (i), so the propositions (i)–(iii) are inconsistent. As Schrenk (2004) puts it,

> A main element of this *reductio* is that speeds combine to average [(iii)], where weights add [(ii)]. This would be impossible however if speed were directly proportional to weight [(i)]. (Schrenk 2004: 6)

As Brown (2000: 529) bluntly puts it, 'That's the end of Aristotle's theory.'

The first question which should be asked is, of course, on what grounds (i), (ii), and (iii) count as (part of) 'Aristotle's theory'. When such a claim is made, there is an implicit claim that the propositions are historically relevant *as a group*, for example because some relevant figure presented them as (part of) a group of propositions at the time. In addition one might expect the propositions to be scientifically united in a sense which makes the inconsistency interesting or important. These are the questions familiar from previous chapters, but the difference in this case is that these sorts of questions have already been asked by authors such as Gendler and Schrenk. In particular it has been asked on what grounds propositions (ii) and (iii) should be considered 'part of the theory'.

This case is particularly striking. It's bad enough if the propositions in question are not 'pointedly grouped' in a way which makes the inconsistency interesting or important. But in this case one can question whether assumption (iii) has any historical relevance at all. First consider assumption (ii). It is clear enough that this assumption is historically relevant in a certain sense: in various different contexts it was taken for granted that if one weight is combined with another the final weight is the sum of the two original weights. This clearly wasn't an assumption specific to Aristotle's theory of motion, but it *was* an assumption believed to be true by a significant number of the relevant individuals. And it clearly unites with assumption (i) as part of an investigation into the motion of combined bodies. Whether such an investigation was considered before Galileo is not at all clear, but at least there *was* a good reason to put (i) and (ii) together (even if nobody before Galileo thought to ask the question).

However, the historical relevance of assumption (iii) is not nearly so obvious. Gendler explicates it thus:

Natural speed is a property such that if a body A has natural speed s_1, and a body B has natural speed s_2, the natural speed of the combined body A-B will fall between s_1 and s_2. (Gendler 1998: 404)

There is no evidence in any of the secondary literature (and in this I include Galileo's works) that the Aristotelians committed themselves to this assumption. Indeed, there is no evidence that Aristotelians even *considered* how 'combined bodies' would fall. But if they simply didn't say anything about such combined bodies, it is obviously somewhat unfair to attribute assumption (iii) to them and then label their position inconsistent—the onus lies with the critic to provide historical evidence that they *did* commit to it in one way or another. In addition, it can hardly be taken for granted that Aristotelians would have signed up to (iii) on the grounds that its truth is 'just obvious', as might be thought of weight additivity (cf. Gendler 1998: 415). Unlike weight additivity, speed mediation is not something one comes across with regularity in everyday life.[1]

But clearly without assumption (iii) the inconsistency claim falls down. Faced with Galileo's argument, the Aristotelian can simply say that when two objects are combined they fall at a natural speed proportional to the combined weight. This is a little awkward, since if there are really two *different* objects, then each ought to fall at its respective natural speed, in line with (i). But any lingering hint of inconsistency disappears if the Aristotelian takes a stance on whether a thing is *really* one object, or *really* two objects. That is, if two objects are combined in such a way that they become one object (e.g. two snowballs pushed together to form one), then they will fall at a speed proportional to the combined weight. But if two objects are combined in such a way that they remain two distinct objects (e.g. two snowballs linked by a piece of string), then they will fall at their respective natural speeds.

Clearly the Aristotelian is now going to end up with conceptual and/or empirical difficulties. She has to say that there is a fact of the matter as to whether a body counts as one thing or more than one thing. If we imagine a

---

[1] Gendler (1998: 415), drawing on Mach, attempts to cash out what is meant by 'just obvious' here in terms of 'tacit knowledge', 'stores of unarticulated knowledge of the world', 'background beliefs', and 'information about the world which was, in some sense, there all along.' The point is that there are some things you just can't help but believe, including weight additivity (because we all have experience of picking up different numbers of things). But we don't have similar experience of speed mediation or non-mediation.

bag of 100 frozen peas dropped from a plane flying over the Sahara (say), then at first we will have one object falling very quickly, and then it would seem that there will be a point at which the peas have thawed and they suddenly reduce their speed to 1/100th of the original, since then we consider the weights of the individual peas. Of course there is no *logical* difficulty here, but rather (at worst) an empirical conflict. If troubled by such examples, the Aristotelian could claim that entification—the number of bodies present—is a matter of degree (see Gendler 1998: 406 for more on this option).

The most important point for my purposes is that the Aristotelian was *free to choose* which of these routes to take, because at the time of Galileo she hadn't committed one way or another. If those making the inconsistency claim believe otherwise—that there *is* a clear commitment to the inconsistent assumptions in the real history of science—then it is up to them to provide the relevant textual evidence. More likely in this case, as in the case of the early calculus, the inconsistency claim is meant to be in the spirit of a 'rational reconstruction' of science, that somehow 'captures' what really happened. Again, such talk should be dismissed as empty until the critic in question has explicated what is meant by 'captures'. Without a good story to tell on this front, to speak of an 'inconsistency in science' or an 'inconsistent scientific theory' here is unhelpful and dangerously misleading. Aristotelians were not inconsistently committed in the suggested way, and Galileo's thought-experiment does not establish that they were.

7.3 Olbers' Paradox

Olbers' paradox is an old problem in cosmology. In modern spacetime theories the problem disappears, but some have claimed that it ought to have been a serious problem for scientists up until certain advances in cosmology were made in the twentieth century (not unlike Newtonian cosmology). But before the twentieth century the problem was not given much attention; this is why Jaki (1969) refers to the 'paradox of Olbers' paradox'. In hindsight, at least, Olbers' paradox is fairly obvious, and Jaki spends most of his book trying to understand how relevant individuals could have failed to appreciate its force. It is usually referred to as a 'paradox' as opposed to an inconsistent theory, but certainly it is paradoxical because of the apparent inconsistency of certain appealing assumptions.

The basic problem here is that, given fairly appealing assumptions, it seems to follow that the night sky should be extremely bright and not dark, as it is. In fact the considerations are very similar to those we saw in Chapter 5 in the case of gravity. The basic point is that the combined effect of the light reaching us from all the stars in the universe should mean that they add up to create an extremely bright night sky. In the case of gravity we saw that the effects of bodies in different directions would cancel each other out, because of their vectorial nature. But, since it is a scalar quantity, this wouldn't happen in the case of light intensity. Olbers, writing in 1823, calculated that the light at any point should be infinite, because it would be equal to $n\frac{\delta^2}{4}(1 + 1 + 1 + 1 + \ldots)$ (Jaki 1969: 135). However, because some stars would get in the way of the light of other stars, in the end the night sky should not be infinitely bright, but still at least as bright as in the daytime.

The first thing to notice here is that we only reach inconsistency when one adds to the relevant cosmological assumptions (to which I'll come in a moment) the assumption that the night sky is dark. Even on a popular conception of *theory* it is perfectly natural to place this assumption outside 'the theory', and to say not that the theory is inconsistent, but that it is inconsistent with what is observed. However, in this book the importance of an inconsistency is not being judged by what is considered to be internal and external to 'a theory'. Instead what matters is the historical relevance of the propositions, individually and as a group, and the character of the grouping. So it doesn't matter for present purposes that the assumption 'the night sky is dark' does not usually count as 'part of the theory': there might still be an interesting and important inconsistency here, with important lessons for the philosophy of science.

However, there is some doubt that an historically relevant set of propositions can be put together which truly do entail a bright night sky, even on a generous understanding of 'entailment'. The basic assumptions are as follows:

(i) The universe is infinite in age and size.
(ii) The stars as we see them go on forever (the star distribution is, at some scale, homogeneous).

But it should be fairly obvious that these assumptions by themselves do not entail that the night sky is as bright as the day. Olbers, for example, explained away the 'problem' by assuming that matter in the universe (such as gas clouds) blocks much of the light. Now, of course, in the latter stages of the nineteenth century thermodynamics spoke against this 'blocking' possibility. It became clear at that time that it would be impossible for something to continuously block light without emitting any of its own. But this fact does nothing to prevent Olbers and others from avoiding inconsistency by assuming such a thing *is* possible. And of course this is not the only possibility: others supposed that there must be gaps in the ether which light simply can't get through, or that space might be non-Euclidean in a way which prevents the paradox. These latter two options were clearly available after Olbers' preferred explanation became untenable in the latter decades of the nineteenth century.

So in other words we only get inconsistency if we add to propositions (i) and (ii) other propositions such as the following:

(iii) The night sky is dark.
(iv) Nothing can block light without re-emitting it.
(v) There are no gaps in the ether.
(vi) The universe is Euclidean.

Now clearly this set of propositions isn't historically relevant in any important, scientific sense. If anybody ever committed to all of these assumptions at once it was a single individual working alone, not a community of scientists. And anyway, it's contentious at the least to claim that propositions (i)–(vi) are inconsistent. For example, according to Jaki (1969: 228f.) William D. MacMillan suggested at one point that light might actually change into matter under certain circumstances. Strictly speaking, to reach inconsistency one would have to add to our list of propositions the negation of all such radical possibilities.

In this case we find that, once again, there is no significant, historically relevant inconsistency. But this case is less significant than most because one requires a large set of propositions to achieve inconsistency which quite clearly would never have been jointly committed to. This time it is not that the unifying question is not asked (as in the cases of the later Bohr theory, Newtonian cosmology, and Aristotle's theory of motion). Many asked the question 'What should the brightness of the night sky be, according to what we believe about the universe?' Olbers certainly wasn't the first: Kepler asked it in 1610, and Halley and Cheseaux followed in the eighteenth century (it is an historical accident that we call it *Olbers'* paradox). But nobody who asked the question thought that there was a serious problem, either because they asked the question before thermodynamics was on the scene, or for one of many other reasons. Seeing an inconsistency in the real history of science in this case represents, as Jaki (1969: 235) puts it, 'an utter disregard of scientific history.'

7.4 Classical Electrons

In Chapter 4 I noted that introducing a self-force into the Lorentz force equation (LFE) in classical electrodynamics (CED) causes all sorts of problems. We saw that Belot prefers to see the LFE in the form $\mathbf{F}_{tot} = q(\mathbf{E}_{tot} + \mathbf{v} \times \mathbf{B}_{tot})$, but that the self-field part of this equation is useless unless one introduces a model of the charged particle in question. Belot (2007) introduces a model of a rigid, spherical charged body with 'continuum many parts', but if one introduces relativity such a model is physically impossible. Of course, in the quantum era charged particles such as the electron are conceived in a way completely foreign to the classical ideas discussed in Chapter 4. But if we ask the question how scientists in the early decades of the twentieth century conceived of charged particles, especially the electron, a potential inconsistency looms.

Mathias Frisch (2005a, 2008) argues that, whether one conceives of electrons as extended or as points, one encounters difficulties—first encountered in the early twentieth century—which have never been fully overcome. To this day, 'no extended-particle model that has been explored in any detail is fully relativistic' (Frisch 2008: 104). And if a particle is a point, one has a very hard time avoiding—in a physically realistic way—the conclusion that the particle will have infinite energy and mass. Frisch (2008) refers to 'conceptual problems in classical electrodynamics', and states that 'there is a tension between the notion of a discrete particle and relativistic field theories' (Frisch 2008: 104). And Frisch has also suggested that this 'tension' is in fact correctly described as an inconsistency (although he has also said that the inconsistency presented in Chapter 4 'locate[s] the inconsistency more precisely' (Frisch 2005a: vii)).

In fact, there is plenty of literature dating back at least to the 1960s which describes the noted 'tension' as an inconsistency at the heart of CED. Landau and Lifshitz labelled the theory inconsistent in 1962, Feynman was quick to follow in 1964, and Shapere gave a more philosophical discussion of the inconsistency in 1969. The relevant passages from Landau and Lifshitz, and Feynman are as follows:

Since the occurrence of the physically meaningless infinite self-energy of the elementary particles is related to the fact that such a particle must (because of the impossibility of rigid bodies, according to special relativity) be considered as pointlike, we can conclude that electrodynamics as a logically closed physical system presents internal contradictions when we go to sufficiently small distances. (Landau and Lifshitz 1962: 102. See also p.44)

The concept of simple charged particles and the electromagnetic field are in some way inconsistent. (Feynman et al. 1964: 28–1)

And Dudley Shapere draws on Landau and Lifshitz, writing,

[W]hereas Lorentz's theory *precluded* the electron from having a zero radius, the theory of relativity *requires* that it have this characteristic.... Classical (relativistic) electrodynamics thus appears to contain a contradiction. (Shapere 1969: 142f.)

If we denote as (Z) the statement 'the electron has zero radius' then from core elements of CED it follows that $\sim$(Z), and from special relativity (SR) it follows that (Z), so that from relativistic CED it follows that (Z)&$\sim$(Z).

One doesn't need to ask what CED 'really is' here. The only relevant posit Shapere makes use of concerns the electrostatic energy of a charged particle. He expands his reasoning as follows:

[T]he electrostatic energy of a charged sphere of radius r and charge e is (except for a numerical factor) equal to e^2/r; this formula implies that a charged sphere of zero radius would have infinite energy, or, if we apply the Einstein relation $E=mc^2$ between energy E and mass m, infinite (rest) mass. However, the electron does not have infinite energy or mass. (Shapere 1969: 138f.)

Therefore, by modus tollens, the sphere does not have zero radius, so $\sim$(Z). Shapere establishes (Z) by drawing on Yilmaz:

An elementary particle is, by definition, a material object which takes part in physical phenomena only as a unit. In other words, from the physical point of view it should not be useful to think of any component part of an elementary particle or to analyze it further. In order to describe the state of the motion of an elementary particle, it is sufficient to know only its position, velocity, and rotation as a whole. It is clear that this would imply a rigid structure if the particle had any classically meaningful extension at all. Thus, elementary particles must be pictured as point particles in the theory of relativity. (Shapere 1969: 142)

So *if* electrons are elementary particles (we will explore this 'if' below), we have (Z)—electrons have zero radius—and thus our contradiction.

Shapere's reasoning is quite persuasive, and many have been convinced by the inferences he makes. Now of course there *is* an inconsistency here—as we have seen throughout the previous chapters there is always an inconsistency if one really wants to find one. But if we eliminate *theory* and identify the full set of inconsistent propositions, what is there to say about the historical relevance and overall significance of the inconsistency?

7.4.1 *Reconstruction of the inconsistency*

First consider again the way in which Shapere moves from CED to the conclusion that the electron cannot have zero radius ~(Z). We have,

(i) The electrostatic energy of a charged sphere of radius r and charge e is (except for a numerical factor) equal to e^2/r. [From CED]
(ii) (Z), or in other words r = 0. [Assumption for reductio]
(iii) $E = mc^2$. [From SR]
(iv) A charged sphere of zero radius has infinite energy and mass. [From (i), (ii), and (iii)]
(v) An electron does not have infinite energy or mass. [Fact]
(vi) ~(ii), or in other words ~(Z), or in other words ~(r = 0). [RAA on (iv) and (v)]

Now, as it stands this doesn't *quite* work. Ignoring very minor niggles, the most important gap in the derivation is the move from (i), (ii), and (iii) to (iv). Really (iv) should read 'A charged sphere of zero radius has infinite *electrostatic* energy and mass'. This then allows for the possibility that it also has mass of a non-electrostatic kind which is infinite and negative, and which cancels the electrostatic mass to give a sensible, finite, net mass (similarly for energy).

This approach, known as renormalization, has been finely tuned and continues to play a role in current theories of the electron (cf. Muller 2007, §6). However, I am not here considering the role played by CED in *current* science: the question I want to ask is whether there was a genuine inconsistency in the early twentieth century, when CED was seriously committed to as a 'candidate for truth'. In fact at that time the electron simply wasn't considered to be a 'point', and the suggestion that it could be a point, and that the problem of infinities could be handled by an infinitely negative 'bare' mass, was some way off (Dirac in the 1930s). Now, to make the inference go through one really needs to add the proposition 'An electron cannot have an infinitely negative, bare mass', and there was no *explicit* commitment to this proposition by the relevant individuals in the relevant period of history. But nevertheless, it makes sense to include this as an implicit assumption. After all, there was much opposition to this idea even in the 1930s, a time when physicists were increasingly open to unintuitive posits (given surprising developments between 1900 and 1930).

Further evidence that scientists were implicitly committed to there being no infinite and negative 'bare' mass is that, during this period, all of the discussion was about *extended* electrons. Why weren't point electrons being considered? Part of the story here has to be that point electrons just don't make intuitive sense. And part of the reason for *this* has to be that the self-field strength at the point of the electron would have to be infinite. We don't have to find the relevant scientists *explicitly* making this argument for us to reasonably attribute the relevant doxastic commitment to them. Their obvious aversion to point electrons is evidence enough.

Since there is ample evidence that physicists at the time agreed that electrons could not be points, the question which remains is whether they were *also* somehow committed to electrons actually being points, and having a zero radius. Following Shapere's account, we might move from SR to the claim that the electron has a zero radius (Z) in the following way:

(i) An electron is an elementary particle.
(ii) ~(Z) or, in other words, one side of an electron is some finite distance from the other side of the electron. [Assumption for reductio]

(iii) Any phenomenon affecting one side of an electron equally affects the other side of the electron at that same instant of time. [From (i) and (ii)]
(iv) Information travels at a finite speed. [From (SR)]
(v) ~(iii). [From (ii) and (iv)]
(vi) ~(ii), or in other words (Z). [RAA on (iii) and (v)]

Now, this derivation is sound enough for me; nothing relevant is gained if we make explicit any implicit assumptions necessary to satisfy the most fastidious logician. However, absolutely crucial for the derivation is proposition (i): 'an electron is an elementary particle'. Is this proposition somehow included in CED or SR, or is it rather smuggled in as an extra assumption? Probably the vast majority of us would agree that CED and SR are independent of this proposition.[2] The conclusion is that Shapere might more accurately have said not that classical (relativistic) electrodynamics is inconsistent, but instead that this theory is inconsistent *with* the claim that the electron is elementary (especially since Yilmaz, on whom Shapere draws, makes reference not to 'electrons' but only to 'elementary particles'). But of course, given theory eliminativism, I don't want to be fussy about whether or not some 'theory' is inconsistent. The question which really matters is whether the relevant assumption set is historically relevant and reveals a significant and important inconsistency in science.

7.4.2 *What lessons?*

The inconsistent assumption set in question is as follows:

(1) The formula for the electrostatic energy of a charged particle (assumed to be a component part of Maxwell–Lorentz CED)
(2) SR
(3) If an electron has infinite *electrostatic* energy/mass, then it has infinite energy/mass (an electron cannot have mass of a non-electrostatic kind which is infinite and negative)
(4) Electrons are elementary

---

[2] Recall Belot's statement to this effect from Chapter 4: '...one is free to stipulate a notion of particle for a given investigation'.

These propositions are inconsistent, but did anybody ever jointly commit to them? That depends on what you mean by 'commit', but first let's consider the possibility of a genuine doxastic commitment in the history of science, back in the early twentieth century. Certainly at this time the relevant part of CED—concerning the electrostatic energy of a charged particle—was considered a serious candidate for the truth. And I have already allowed that assumption (3) is reasonably considered an implicit commitment. With SR one has to be careful. It is usual to hear that it was 'introduced by Einstein in 1905', but there was hardly universal assent to the idea from Day One. In fact, when one looks to the commitments of the most important relevant figures in the years 1905–20, things get messy very quickly. A whole chapter could easily be spent on this issue, but one can get a good flavour of events if we consider just a few of the relevant individuals.

Let us ask the question who might have committed to all of the propositions (1)–(4) given above. Consider Max Abraham first. He certainly committed to (1), and in the first few years of the twentieth century he introduced a model of the electron as a rigid sphere with a uniform surface or volume charge distribution. In a 1904 paper (responding to Lorentz) he stated explicitly that one of his assumptions was that 'A change of form of the electron is not possible' (Goldberg 1970: 19). So we can safely say that Abraham committed to (4). All that remains, then, is to establish that Abraham committed to (2), but it's very clear that he didn't. In fact, he *never* accepted relativity, right up to his death in 1922 when for virtually everybody else there was no longer any doubt about it (see e.g. Kragh 1999: 113). Thus Abraham was not committed to (1)–(4). Indeed, it is worth noting that Abraham was well aware that he couldn't commit to a rigid electron *and* relativity; this was obvious to everybody in the community at least as early as 1906 (see Kragh 1999: 110). In other words there was *never any chance* of him committing to all of (1)–(4); the inconsistency was too obvious for that. Walter Kaufmann can also be mentioned here. His experiments in 1906 were seen by many to go strongly against relativity; at a major meeting in Stuttgart in 1906, '[M]ost of the participants were on Kaufmann's side, for the rigid electron and against the deformable electron and the relativity principle' (Kragh 1999: 112).

The main rival to Abraham's model of the electron in the early years of the twentieth century was Lorentz's model. Here the electron was a sphere with a uniform surface charge distribution when at rest, but contracted into an ellipsoid when in motion. This allowed for a commitment to relativity, and indeed Lorentz's theory was sometimes referred to as the 'Lorentz–Einstein theory' (Kragh 1999: 112). This really only leaves proposition (4) in question; did Lorentz think that electrons were elementary?

Well, he certainly didn't think of the electron as a point, or as a rigid body. Indeed, in his 1909 book *The Theory of Electrons* he repeatedly refers to the 'parts' of an electron, and designs equations on the basis of the electron having parts, which themselves can experience forces. To give just two examples, in §32 of the book he writes that 'we can speak of forces acting on its parts' and 'the parts of an electron cannot be torn asunder by the electric forces acting on them' (Lorentz 1916: 43). And in §181 of the book he suggests that the electron 'has the properties of a very thin, perfectly flexible and extensible shell, whose parts are drawn inwards by a normal stress' (Lorentz 1916: 213f.).

At one point (§182) Lorentz does consider that 'perhaps, after all, we are wholly on the wrong track when we apply to the parts of the electron our ordinary notion of force' (Lorentz 1916: 215). But this just goes to show that Lorentz was, for the most part, thinking in terms of forces affecting parts of electrons. With this commitment in place, Lorentz is clearly not committed to proposition (4), and therefore not inconsistently committed to all of (1)–(4).[3]

Now there *may* be a sense in which he was inconsistently committed to a slightly different set of propositions (Frisch 2005b: 674). One thing Lorentz needed for his theory was an account of why the parts of the electron didn't fly apart due to their mutual electrostatic repulsion. When Poincaré introduced in 1905 a non-electromagnetic force (the 'Poincaré counter-pressure') holding the parts of the electron together, Lorentz tentatively adopted it (Frisch 2005b: 665). But Lorentz always believed that such a 'counter-pressure' could not ensure the stability of an *accelerating* electron, and yet he also naturally assumed that electrons *do* accelerate and *are* stable. Frisch (2005b) goes on to argue that, given this problem, Lorentz did not believe that his theory of the electron could be true. Instead, Lorentz

---

[3] If one avoids the inconsistency by conceiving of the electron as having parts, then one might object that the same problem arises for the parts, and the parts of the parts, and so on ad infinitum. But there are various options here if one is metaphysically sophisticated. Most obviously one might think of an electron as consisting of a continuum of parts, each with an infinitesimal charge. One might even suggest that the electron's parts are not charged, and that the electron's charge is somehow 'emergent'.

(§182) suggested that it was 'legitimate to maintain the hypothesis of the contracting electrons, if by its means we could really make some progress in the understanding of the phenomena' (Lorentz 1916: 215). In other words Lorentz's commitment was 'partly instrumental' (Frisch 2005b: 678). Clearly in such circumstances there is no doxastic commitment to the inconsistent propositions; instrumental or pragmatic commitment will be considered shortly.

Lorentz was actually typical in thinking in terms of 'parts' of an electron in those years. Bucherer and Langevin independently constructed a different model of the electron in 1904, similar to the deformable electron of Lorentz but with invariant volume under motion (Kragh 1999: 107f.). As with Lorentz, the deformability of the electron tells us that these individuals were not committed to electrons as elementary (at least not in the sense required to make (1)–(4) inconsistent). Further, Bucherer was not even committed to Einsteinian relativity, even in the 1920s (Kragh 1999: 113). And Poincaré is another relevant individual who rejected both (2) and (4). The introduction of the Poincaré counter-pressure speaks to his rejection of (4), and Miller (1981 225) discusses his attitude towards Einsteinian relativity.

What of Einstein himself? He was more interested in developing relativity as a purely kinematical theory, silent on the question of the hidden nature and structure of electrons (see e.g. Kragh 2001: 211ff.). But if pushed he would have committed to extended electrons over point electrons, since he was well aware that point-electrons would have infinite energy according to CED. As Mc Cormmach (1970: 62f.) notes, comments Einstein made in later years indicate that this was Einstein's thinking at the time. And it is obvious that Einstein would have considered these extended electrons deformable bodies, given relativity. Hermann Minkowski's attitude is representative of those committing to relativity at that time: '[A]pproaching Maxwell's equations with the concept of the rigid electron seems to me the same thing as going to a concert with your ears stopped up with cotton wool' (Minkowski in 1908, cited in Kragh 1999: 107).[4]

---

[4] Of course, many other historical figures could be discussed here. Gustav Mie developed a theory between 1911 and 1913, with electrons conceived as 'knots of the electric field in the ether' (see Kragh 1999: 117). But Mie avoided inconsistent commitment by making changes to the EM equations at play (Kragh 1999: 117). J. J. Thomson considered the electron to be extended (even if he sometimes referred to it as a 'point'), and Mendeleev considered the whole question of the structure of the electron to be 'the province of fancy and not of science' (Kragh 2001: 206–9).

One may insist that there must have been a time when relativity was widely established, CED was still considered a serious candidate for the truth, and electrons were considered elementary particles. But when was there widespread commitment to electrons as *elementary* particles? Pais (1972: 79) writes that, 'In the first quarter of this century, still the classical period, electrons were pictured as small but finite bodies...[T]he main question became how to reconcile such a particle picture with special relativity theory.' This is a bit misleading, but only because the question of electron structure gradually came to be ignored altogether in the first quarter of the twentieth century (see Kragh 2001: 215). That is, physicists were increasingly silent on the question of the 'real' nature of the electron. They considered it a 'pseudo question' (Kragh 1999: 115), adopting an 'antimodel attitude' (Kragh 2001: 216), and simply treated the electron as a point for convenience (e.g. Bohr 1913a). Only in the 1920s did physicists start to explicitly argue for electrons as points (especially Frenkel in 1925; see Kragh 2001: 216). But by then any *doxastic* commitment to all of (1)–(4) would be highly unlikely given non-classical (quantum) developments.

The simple story here is that the inconsistency of (1)–(4) is extremely obvious, and so it should be obvious that no commitment to *all* of (1)–(4), at the same time, as 'candidates for the truth', will be found in the history of science (cf. the case of the early calculus). To imagine that this could be the case would be to do an injustice to practicing scientists. Any scientists committing to three of the propositions (1)–(4) found a way to avoid commitment to the fourth precisely because of the inconsistency. Dirac, for example, presented the electron as an elementary point in 1938, with infinite electrostatic energy. But of course Dirac explicitly rejected (3), introducing renormalization in order to make this possible (cf. Muller 2007: 268–9).

This isn't to say that there wasn't some *other* type of 'instrumental' or 'pragmatic' commitment to all of (1)–(4) from time to time. For example, in the 1920s and 1930s (and even up to the present day) CED was (is) still *used*, at least, in many contexts. But the significance of the inconsistency is reduced, since as we have seen in several of the previous examples there can be good reason to *work* with a set of inconsistent propositions. And the 'threat' of ECQ turns out to be no threat at all as previously explained; the story is not significantly different from that already detailed in Chapters 3 and 4, and I will not repeat myself here.

One peculiarity remains for this example. I've said that as physics developed in the twentieth century the debate over classical electron models disappeared, replaced by quantum considerations and a commitment to electrons as elementary. As Dirac put it in a famous statement from 1938, '[T]he electron is too simple a thing for the question of the laws governing its structure to arise' (1938: 149). But in fact the question of classical electron models never fully disappeared. Indeed, one finds detailed discussion of classical electron models in Yaghjian (1992), Spohn (2004), and Rohrlich (2007), to give just three prominent, recent examples (see Muller 2007: §§6 and 7, and Frisch 2008, for further details and references). But what would be the point of investigating classical electron models if we are certain that they are not candidates for the truth, or even approximate truth?

The answer ought to be that such investigations are nevertheless illuminating in *some* way. Now, it is sometimes the case that mathematical and scientific investigations proceed without it being clear what payoff, if any, there will be. Such investigations can be justified by pointing to a large number of such investigations which ultimately resulted in a very large, unexpected payoff (e.g. the application of matrix algebra to quantum mechanics). The question for present purposes is what sort of commitment scientists are making to their posits as they develop classical electron models. Curiously, despite no possibility of doxastic commitment, scientists act *as if* they are doxastically committed by carefully avoiding any inconsistent commitments (including commitment to (1)–(4)). The guiding question appears to be 'what *would* the world be like, in reality, if it were classical?' Whether this is a worthwhile question for theoretical physicists I leave for the reader to decide.

7.5 Kirchhoff's Theory of Diffraction

The subject matter for this final case study is the diffraction of light at an aperture. In 1882 Kirchhoff wrote a paper in which he asked the question,

(K) What intensity of light will we find at a given point beyond an aperture which is illuminated by a monochromatic source?

Putting together certain assumptions about the behaviour of light, he was able to derive a formula which provided remarkably successful predictions. This obviously encouraged serious commitment to Kirchhoff's theory. But it turns out that in addition to Kirchhoff's formula, the propositions also entail a contradiction. The case has barely been discussed in the philosophical literature, but it has been discussed in some detail in the relevant *scientific* literature. As Heurtley (1973) writes, 'A problem of continuing interest in scalar diffraction theory is why the mathematically inconsistent theory of Kirchhoff predicts results that are in substantial agreement with experiment' (1973: 1003). I first detail the inconsistency in the next section, before addressing Heurtley's question in §7.5.2.

7.5.1 Reconstruction of the inconsistency

Kirchhoff wanted to explain the distribution of light intensities which are detected beyond an aperture illuminated with monochromatic light. For the sake of simplicity he considered an idealized setup represented by Figure 7.1. Crucial features of the assumed scenario which aren't represented in the Figure are that the screen is infinitely thin and infinitely opaque. His assumptions were as follows:

(i) The light at the aperture behaves just as if the screen were not there.
(ii) The light source emits spherical, monochromatic waves of light.

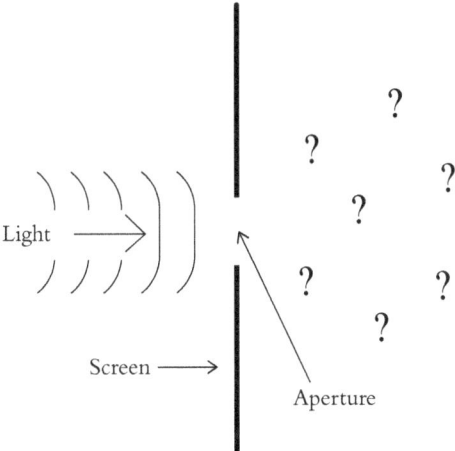

Figure 7.1 Kirchhoff's challenge: How does the light behave beyond the aperture?

210 UNDERSTANDING INCONSISTENT SCIENCE

(iii) The Helmholtz–Kirchhoff integral theorem (a result in mathematics).
(iv) The amplitude of light and the derivative of this amplitude are zero immediately behind the screen.
(v) The Sommerfeld radiation condition.

The main lesson is as follows. From (i)–(v) Kirchhoff was able to derive a startlingly successful formula for the behaviour of light in experiments which come close to recreating the idealized setup of Figure 7.1. But one can also derive a contradiction from these propositions.

Now for the juicy details.[5] Kirchhoff makes essential use of a result in mathematics which says that the magnitude of a given parameter at a point of interest can be determined from knowing the magnitude of that parameter at all points on a surface surrounding that point of interest. This is called a 'boundary value technique', because from knowing what is happening *at* the boundary one can determine what is happening at points of interest

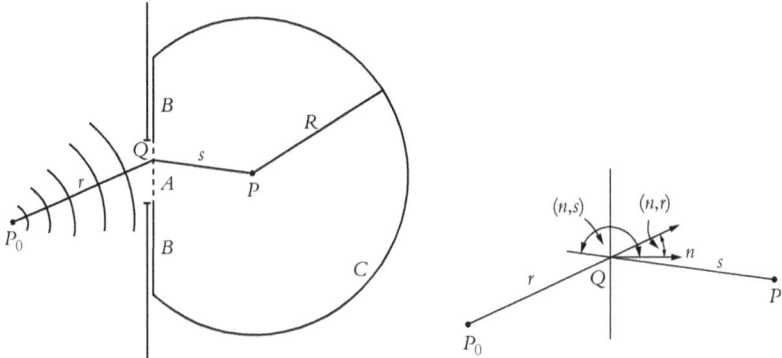

Figure 7.2 Diagrams showing the relevant parameters Kirchhoff's derivation makes use of. P_0 is the source of the light, and P is the point beyond the screen at which we want to know the light intensity. In addition Q is a point in the aperture whose contribution we are considering at a given time, r is the distance from P_0 to Q, and s is the distance from Q to P. An imaginary surface of integration S is comprised of A (the aperture), B (part of the screen), and C (part of a circle of radius R which has P at its centre). n is a normal to the aperture, (n, r) is the angle between this normal and the line joining P_0 to Q, and (n, s) is the angle between this normal and the line joining Q to P
Source: Figure taken from Born and Wolf (1999: 421).

[5] Here I follow Born and Wolf (1999: ch. 8).

inside the boundary. In the present context Kirchhoff's strategy is to determine what is happening at a given point of interest *P* beyond the screen by first determining what is happening at every point on a certain imaginary surface surrounding *P*. As seen in Figure 7.2, the surface he considers is made up of the aperture *A*, part of the screen *B*, and part of a large circle *C* which has *P* at its centre.

First Kirchhoff draws on assumption (iii) to describe the boundary value problem in question. This gives the intensity *I* at a given point *P* indirectly, since it gives us the amplitude *U(P)*, and the intensity and the amplitude are related by the equation $I(P) = |U(P)|^2$. The equation for the amplitude is,

$$U(P) = \frac{1}{4\pi}\left[\iint_A + \iint_B + \iint_C\right]\left\{U\frac{\partial}{\partial n}\left(\frac{e^{iks}}{s}\right) - \left(\frac{e^{iks}}{s}\right)\frac{\partial U}{\partial n}\right\}dS \quad (7.1)$$

Now he draws on assumption (iv) to eliminate the integral over *B*, since according to that assumption *U* and $\partial U/\partial n$ are zero there, so (as can be seen by inspecting equation (7.1)) the contribution to *U(P)* will be zero. Next he draws on assumption (v) to eliminate the integral over *C*, since if one takes the radius *R* large enough it can be shown that the contribution to *U(P)* will be zero (see Born and Wolf 1999: 422). This means that we are left with the equation,

$$U(P) = \frac{1}{4\pi}\iint_A \left\{U\frac{\partial}{\partial n}\left(\frac{e^{iks}}{s}\right) - \left(\frac{e^{iks}}{s}\right)\frac{\partial U}{\partial n}\right\}dA \quad (7.2)$$

Now Kirchhoff draws on propositions (i) and (ii) to describe the behaviour of the light at the aperture:

$$U = \frac{Ke^{ikr}}{r} \quad (7.3)$$

$$\frac{\partial U}{\partial n} = \frac{Ke^{ikr}}{r}\left[ik - \frac{1}{r}\right]\cos(n,r) \quad (7.4)$$

Here *k* is the wavenumber given by $k = 2\pi/\lambda$, where λ is the wavelength of the light, and *K* is a constant. Plugging these into equation (7.2) a straightforward calculation yields,

$$U(P) = -\frac{iK}{2\lambda} \iint_A \frac{e^{ik(r+s)}}{rs} [\cos(n,r) - \cos(n,s)] dA \qquad (7.5)$$

This is Kirchhoff's diffraction formula.

The most obvious manifestation of the inconsistency of propositions (i)–(v) was noted by Poincaré in 1892, ten years after Kirchhoff's paper (Poincaré 1892: 187). The problem is that equation (7.5) disagrees with the boundary assumptions: if one considers what it says about light at the aperture it conflicts with assumption (i), and if one considers what it says about light immediately behind the screen it conflicts with assumption (iv). So it would seem that from propositions (i)–(v), it is possible to derive a proposition saying 'A&~A', where options for 'A' include assumption (i) and assumption (iv).

7.5.2 *What lessons?*

Is this an interesting or important inconsistency from the point of view of philosophy of science? Obviously propositions (i)–(v) are, in the terminology of Chapter 2, pointedly grouped propositions. They are scientifically unified, since they come together to derive an extremely successful formula for the prediction of light intensities in experimental setups similar to Figure 7.1. In addition they are obviously historically relevant, as a group, since Kirchhoff put them together himself and, given the success of equation (7.5), others followed suit. So it might seem that we have a really serious inconsistency here. But we can push harder, asking the question *what is it* that makes the inconsistency important? What exactly do we learn?

I have argued that there are two main ways in which an inconsistency is usually thought to be significant: (i) when scientists make a doxastic commitment to the propositions in question, and (ii) when scientists reason with the propositions in question, but nevertheless avoid ECQ and trust at least some of their inferences. Taking (i) first, it is clear that scientists would never have believed all of the propositions (i)–(v), or even believed them to be candidates for the truth. The reason is that Kirchhoff is considering an idealized setup which couldn't possibly exist in the real world: no screen can possibly be infinitely thin (whatever that could mean) and infinitely opaque. So any doxastic commitment made to the propositions would have included approximation clauses. But what if we 'internalize the approximation' (as we did with Bohr's postulate P2 in §3.2.3.2)? Then we will have serious doxastic commitment to the adjusted propositions. Does the inconsistency remain?

What would it mean to internalize the approximation in this case? Well, we do know that screens can at best be *very* thin and *very* opaque. For an experimental setup with such a screen assumptions (i) and (iv) would be approximately true; in fact the non-zero width of the edges of the aperture would affect only these assumptions. Internalizing the approximation we reach,

(i★) The light at the aperture behaves *approximately* as if the screen were not there.

(iv★) The amplitude of light and the derivative of this amplitude are *approximately* zero (and definitely not exactly zero) immediately behind the screen.

Add these two propositions to (ii), (iii), and (v) and we have a set of propositions (i★)–(v) which Kirchhoff would have put forward as *serious candidates for the truth*. So, again, does the inconsistency remain?

In fact deciding this is a hopeless task, because of the vague nature of the word 'approximate'; it is far from clear exactly what (i★) and (iv★) mean (recall the discussion in §3.2.3.2). We can make *some* headway. For example, since U and $\partial U/\partial n$ are no longer zero, the following term needs to be added to equation (7.5):

$$\frac{1}{4\pi}\iint_B \left\{ U \frac{\partial}{\partial n}\left(\frac{e^{iks}}{s}\right) - \left(\frac{e^{iks}}{s}\right) \frac{\partial U}{\partial n} \right\} dB \qquad (7.6)$$

This term can't be eliminated now, since neither U nor $\partial U/\partial n$ is zero. But neither can it be evaluated, since the nature of U along B (except for being approximately zero) is left totally mysterious. In addition equation (7.5) needs to be changed, since we can no longer assume that, along the aperture, $U = \frac{Ke^{ikr}}{r}$ (equation (7.3)). But again, since we are given no information as to how exactly equation (7.3) should be changed, we don't know how equation (7.5) will be affected. There might well be some way of spelling out the details of (i★) and (iv★) such that the Poincaré inconsistency disappears, but if this is possible at all it is a mathematical problem far beyond my powers as a philosopher, unfortunately!

Suppose it is actually impossible to eliminate the Poincaré inconsistency in this way, even given the vague nature of 'approximate'. What then?

Well, the case certainly shows that in complex physics and mathematics there can be inconsistencies in our doxastic commitments that are astonishingly well hidden, and that may perhaps never come to light. In such a case our scientific community may harbour inconsistent beliefs without anybody ever knowing it, and even whilst informing the wider world that they have very good reason to believe that the propositions in question are all true (given predictive successes, say—see Saatsi and Vickers 2011 on this point). If there *is* an inconsistency in propositions (i★)–(v), it doesn't affect the science at all: predictions are key. As Andrews (1947: 784) puts it, 'The sole virtue of Kirchhoff's theory of diffraction lies in its correct predictions and not in its false assumptions.' The assumption would have been that the de-idealization would—if it *could* be worked through—only marginally affect the formula, and that would have been the end of it. In short, the inconsistency is dismissed as an artefact of the idealization.

The point of interest for our purposes is that, *if* an inconsistency remains in the de-idealized propositions (i★)–(v), in attempting to make it more serious in one way I have made it less serious in another. That is, any inconsistency remaining in the de-idealized propositions is more serious due to it being—at least potentially—*about the world* (rather than some idealized mental construct). But it is less relevant to real science because nobody was ever interested in the de-idealized propositions, even though if pushed it is the de-idealized propositions scientists would have put forward as (serious candidates for) 'the truth'.[6]

But regardless of doxastic commitment, what of ECQ? What of trusting inferences made using Kirchhoff's assumptions? It is obvious what to say in the case of the de-idealized propositions (i★)–(v): the contradiction, if it exists, would never be reached in practice, so explosion is impossible. In fact, nobody would really care about inferences from these propositions, since despite being candidates for the truth they are just useless for making interesting inferences about diffraction behaviour.

But what of the original propositions (i)–(v) which we know *are* inconsistent, and from which a contradiction can be derived relatively easily (as

---

[6] In fact Marchand and Wolf (1966) showed that by making certain changes to assumptions (i) and (iv) one *can* derive Kirchhoff's diffraction formula (equation (7.5)) and avoid the Poincaré inconsistency. However, the changes made to assumptions (i) and (iv) are quite dramatic, and certainly can't be described as (i★) and (iv★) on any stretch of the imagination. For more details see Saatsi and Vickers (2011).

Poincaré demonstrated)? Here the situation is very much like previous cases we have seen (recall, e.g., §3.2.3.2). Scientists were using propositions at least some of which they knew to be false, and in these circumstances the game changes. They do not blindly trust their inferences, but instead examine what is inferred, checking it empirically and conceptually. This is because scientists know that making truth-preserving inferences from merely *approximately* true propositions can take one to both approximately true conclusions and also to radically false conclusions. They are obviously interested in the approximately true conclusions! If they infer something inconsistent with their original assumptions, they will recognize that this inference has zero scientific value, and stop making inferences. Thus ECQ does not threaten, since to reach an arbitrary proposition one first needs to reach explicit contradictories, and then *continue* making inferences (as explained previously).

Even if we accept that ECQ does not threaten, one may still wonder how it makes sense to trust *any* of the content of Kirchhoff's diffraction formula if one knows that *some* of this content conflicts with one's original assumptions. Since the formula straightforwardly contradicts the original assumptions (i)–(v), shouldn't one reject the whole formula as 'an absurd result'? But this would obviously be to throw the baby out with the bathwater, since we know that much of what Kirchhoff's formula tells us is approximately true. We don't need to simply 'trust' what Kirchhoff's formula tells us about points P far from the aperture, since in practice what it tells us is judged conceptually and empirically. An element of trust does come in occasionally, when one considers a new point P close to a range of other points where one knows the formula to give approximately true results. Then one may feel justified in simply trusting what the formula tells us. This is to a certain extent risky, since one knows that one has reached this conclusion by making inferences from false assumptions, and by reasoning in a very similar way to the way one reaches completely absurd results. However, in practice one reasons that it would be surprising if the formula gave approximately true results for a wide range of surrounding points but a radically false result for this new, neighbouring point of interest P. After all, one assumes it is not luck alone which leads us to the approximately true conclusions, but rather the fact that one started with approximately true assumptions and then made truth-preserving inferences in an attempt to stay as close to the truth as possible. This helps us to provide an answer to Heurtley's question, introduced above. Simply put, it needn't be surprising that 'the mathematically inconsistent theory of Kirchhoff predicts results that are in substantial agreement with experiment.' As explained in previous chapters, a set of inconsistent propositions can all be true-or-approximately-true, and if one makes truth-preserving inferences from a set of true and approximately true propositions it shouldn't be surprising if at least *some* results are approximately true (or even true; recall the Goldbach case discussed in §3.2.3.2).

There is more to say here, especially since it turns out that not all of Kirchhoff's original assumptions *are* approximately true (see Saatsi and Vickers 2011). But note now that we are no longer discussing the sense in reasoning from inconsistent assumptions, but rather discussing the sense in reasoning with assumptions one knows to be false (and at best approximately true). Reasoning with inconsistent assumptions slots into this other debate as a special case, where one happens to know that the assumptions are false not only on conceptual or empirical grounds, but on purely logical grounds.

7.6 Conclusion

This completes our journey through eight of the primary examples of what are often labelled 'inconsistent scientific theories' in the literature. In the introduction to this chapter I asked two questions: (i) Does it really make sense to refer to these final four cases as 'inconsistencies in science', and (ii) Are the cases substantially similar, or are they idiosyncratic such that general claims about 'inconsistency in science' are unlikely to be possible?

Certainly in the first two cases in this chapter—Aristotle's theory of motion and Olbers' paradox—when one looks to the real history it is astonishing that anybody ever claimed that these are genuine cases of 'inconsistency in science'. And even the latter two cases do not appear to be inconsistencies in any particularly significant sense. In the case of the 'classical theory of the electron' the inconsistency is obvious, and was recognized by all the relevant individuals in the relevant period of the history of science. Accordingly these individuals took steps to avoid inconsistent commitments, as discussed in §7.4.2. And in the case of Kirchhoff's theory, although we certainly do have an inconsistency in 'pointedly grouped propositions', the character of the community's commitment to some of the propositions means that the inconsistency is not especially interesting or important. Certainly there was never any doxastic inconsistency here, since scientists obviously knew that no screen could be infinitely thin and infinitely opaque! But in addition ECQ never presented any danger, since scientists were interested in how they could use Kirchhoff's formula to get useful predictions, not in how they could use it to derive absurd results. The absurd results (including contradiction) could sensibly be dismissed as artefacts of the idealizations.

So, adding these further four cases to the previous four, we do indeed find that a careful analysis of the relevant history and science brings us to see that,

(i) Many of the examples of 'inconsistent theories' or 'inconsistent science' in the philosophical literature are not sensibly reconstructed as such at all, and,

(ii) Of the remaining examples, most are not interesting or important inconsistencies from the point of view of 'how science works' and the philosophy of science generally.

There are few remaining examples of significant, genuine inconsistencies: the Pauli inconsistency of Chapter 3 and the (I2) inconsistency of Chapter 5 are perhaps the only examples. Thus my second question—'Are the examples similar enough for us to make substantial, general claims about inconsistency in science?'—cannot really be answered here. Certainly we can compare these two examples, but one might demand more examples before we can make real progress with this question. Thus, in the end, my most significant conclusions in this book may have to be metaphilosophical—about debates in the philosophy of science, about the way philosophy of science is (sometimes) done—as opposed to philosophical—about inconsistency in science.

But this doesn't necessarily mean that there aren't some interesting overlaps between the cases which are worth exploring. Apart from similarities between cases of serious doxastic inconsistency, there are similarities between cases of non-doxastic inconsistency, and other similarities besides. Then there are some interesting similarities at the metaphilosophical level, between different philosophical debates *about* inconsistency in science. For an in-depth analysis of the lessons learnt from these eight case studies I now turn to a final concluding chapter.

8
Conclusion

8.1 Introduction

This completes our journey through eight of the primary examples of what are often labelled 'inconsistent scientific theories' in the literature. We have seen that there are similarities in the examples, but that there are also many important differences which are currently overlooked in the literature. These differences suggest that no overarching 'theory of inconsistent theories' is to be expected. Indeed, there are very few substantial generalizations concerning inconsistency in science that fall out of the case studies. The philosopher of science is typically very keen to make substantial generalizations about science, and has made substantial generalizations about inconsistency in science, including how we do and should reason with inconsistent theories, why we fail to notice, and come to notice, inconsistencies in science, and how we can move forward when faced with inconsistencies in science. These generalizations—at least insofar as they are supposed to apply to my chosen case studies—do not succeed. Of course, the case studies have all (or nearly all) been described as 'theories', and been given theory-names, and this in itself suggests that one *ought* to be able to say something quite general about inconsistencies in science, or at least in scientific theories. But this is really just an illusion, created in large part by the fact that the common word 'theory' is used in all of these different cases. Thus theory eliminativism brings us to appreciate the diversity of science. Many of the inconsistent propositions are 'pointedly grouped', in a sense, but the character of the inconsistency is very different because the 'point of grouping' is different.

I will turn to theory eliminativism—its costs and benefits, and the scope for wider application—in §8.3. But first we may consider, in §8.2, which generalizations *do* fall out of the case studies, even if the lessons turn out to be modest. In §8.2.1, I consider the suggestion that asking particular questions brings inconsistencies in science to the surface. In §8.2.2, I highlight the important distinction between inconsistency and 'mere' implausibility, a distinction which has surfaced in different ways in the case studies. In §8.2.3, I take a look at the different relationships between 'as if' reasoning and inconsistency in science. In §8.2.4, I turn to the difficulties which sometimes prevent us from seeing that a given set of propositions *are* inconsistent. And in §8.2.5 I consider what the case studies add to the logic-driven/content-driven debate introduced in Chapter 1. In §8.4, I offer some final, concluding thoughts.

8.2 Lessons from the Case Studies

There are some general lessons that can be gleaned from the case studies, both regarding how science works and how philosophy of science works. Some of the generalizations concerning science are not very substantial, but that in itself is a result: in certain cases it will be interesting to note just why more substantial generalizations cannot be made, and shouldn't be expected. In the end the differences between the cases are just as revealing as the similarities, if not more so.

8.2.1 *Asking questions*

In nearly all of the case studies a particular scientific question plays an important role in bringing the inconsistency to light. Some relevant questions are as follows:

(1) In the case of the later Bohr theory of the atom, Pauli's question about how crossed electric and magnetic fields affect a hydrogen atom during a certain adiabatic transformation.
(2) In the case of Newtonian cosmology, the question of what the net gravitational force on a given test particle is.
(3) In the case of the early calculus, the question of how to justify the individual moves made within the algorithm.
(4) In the case of Aristotle's theory of motion, the question of how bodies fall when they are combined together in some way.

(5) In the case of Olbers' paradox, the question of how bright the night sky should be.
(6) In the case of classical electron models, the question of the structure and properties of electrons, and how they are affected by incoming forces.
(7) In the case of Kirchhoff's theory of the diffraction of light at an aperture, the question of how light will behave in a particular experimental setup, and in particular how it will behave at the aperture and behind the screen.

In each of these cases there is a sense in which the question is the catalyst which leads to the inconsistency. But is there anything more substantial we can say which will help us to see how the *next* important inconsistency in science will surface?

In fact when one looks to the details of the cases there are important differences which prevent more substantial generalizations. Consider first the case of the later Bohr theory. In asking his question Pauli was led to a conclusion which contradicted a stipulation of the theory, namely, that there could be no 'pendulum orbits'. In this particular case all the propositions were already in place, and merely had to be put together in the right way to reach the contradiction. Kirchhoff's case is similar but also different in certain respects. Both Pauli and Kirchhoff imagined a particular experiment, and considered what accepted science would predict in such circumstances. In both cases the prediction contradicted a claim of the theory. But in the case of Kirchhoff's theory, propositions had to be introduced particular to the experimental setup in question, such as how light would behave at the aperture and behind the screen. These propositions are very specific to the particular example: it is no exaggeration to say that the propositions did not exist at all until Kirchhoff's question was asked. The question actually generates the propositions. In Pauli's case, however, no new propositions were required to answer his question. So despite the similarities between these two cases, there is an important difference which affects the character of the inconsistency: in Pauli's case the inconsistency already existed within propositions being used by the relevant community, but in Kirchhoff's case the inconsistency did not exist before he asked the question, because some of his assumptions did not exist before he asked the question.

Another important difference comes in the fact that Kirchhoff's question pertains to an idealized experimental setup, whereas Pauli's does not. Pauli's assumptions were assumed to be candidates for the truth when he made his calculation, so the derived contradiction suggested a contradiction *in the world*. But the contradiction in Kirchhoff's theory merely told scientists that the *idealized setup* in question was self-contradictory. Whether this contradiction crosses over to the real world when Kirchhoff's theory is de-idealized is anybody's guess: precisely how assumptions (i) and (iv) in the theory should be de-idealized is up for grabs, and whether the inconsistency remains after the relevant changes have been made is unknown. This makes the inconsistency much more tolerable: one is free, in this case, to assume that the inconsistency is an artefact of the idealization. Not so in the Pauli case.

The inconsistency in Aristotle's theory is similar to the Kirchhoff example, although once again different in some important respects. It is similar in that, once we ask a question about what will happen in a very specific circumstance, certain propositions are (potentially) drawn together which turn out to be inconsistent. In Aristotle's theory the question is what speed we should expect two or more bodies of different weights combined together to fall at. This question generates what might be called 'new propositions', as in Kirchhoff's case, although they are not *quite* as new as in Kirchhoff's case. For example, one ends up asking the question whether two objects bound together are still two objects, or are really one, or whether there is no fact of the matter. Galileo shows that if one is not sufficiently careful in answering these questions then one ends up contradicting oneself. However, it isn't clear whether the 'new' propositions required to reach inconsistency were part of the Aristotelian picture or not. Certainly they are the kind of thing the Aristotelian *may* have considered: unlike in Kirchhoff's case, the question whether something is one object or two may well have come up in contexts other than the one in question. But at any rate, the inconsistency is so obvious in the case of Aristotle's theory that a real-life commitment to the relevant propositions is extremely unlikely. In Kirchhoff's case the inconsistency is more deeply buried, and the propositions have desirable virtues such as explanatory and predictive success.

So now we have three cases: one where asking a question exposes an inconsistency within propositions already accepted by the community (Bohr), one where a question naturally gives rise to propositions which, when combined with accepted propositions, are inconsistent (Kirchhoff), and one where a question *could* give rise to propositions inconsistent with other accepted propositions, but where commitment to the new propositions is highly doubtful, not least because the inconsistency is so obvious (Aristotle). These cases alone show that the character of an inconsistency, and the corresponding significance of an inconsistency for philosophy of science, depends on various material facts about the specific case. An analysis which makes inconsistency a yes-or-no matter does not do justice to the history (see §8.3).

Newtonian cosmology is another interesting case insofar as the question is concerned. Clearly noticing the inconsistency of the propositions depended on asking the right question, which Hugo von Seeliger finally did in 1895. The claim was that the question brought together propositions which were already 'established' in some sense, which were already in the community. This would make it something like the later Bohr case, where arguably the inconsistency was implicit in the assumptions of the community for some time but went unnoticed. But if one takes a closer look there are some important differences between the cases. In Old Quantum Theory all of the relevant propositions were being used by the community in the context of asking questions about a certain domain of phenomena, and the question Pauli asked was just the type of question which was already being asked by others. In the case of Newtonian cosmology the relevant propositions were not really being *used* at all for the period 1700 to 1900, and some of them weren't even on the mental horizon of most individuals for most of that time. Especially in the nineteenth century, Seeliger's question wasn't even the kind of question most scientists would take seriously. However, in the eighteenth century, for some relevant individuals at least, it was the kind of thing that would have been considered. So for this period the historical and scientific relevance of the inconsistency is considerably higher. The question may have been asked from time to time (cf. Richard Bentley's correspondence with Newton), and the inconsistency simply not noticed because of its being difficult to expose (see §8.2.4, this volume). So was Newtonian cosmology implicitly inconsistent, and just waiting for the question to be asked which would make people *notice* that? The question is a mistake. In the sense that certain individuals would have seriously committed to the relevant assumptions had they been asked, perhaps yes, but in the sense that certain individuals *did* commit to the relevant assumptions, not really. For the most part the propositions weren't *jointly* committed to: Seeliger's question had to be asked for this to happen. Accordingly, any modern reconstruction which *does* put them together as a group for philosophical analysis needs to be very carefully interpreted. Again, the relationship between the philosophy of science and the history of science is of paramount importance here.

The case of the early calculus has a different character once again. Berkeley can be compared with Galileo here. He asks a given question—this time 'How are the moves made within the algorithm justified?'—and provides an answer which leads to a contradiction. But as in the case of Aristotle's theory of motion, it is far from clear that the given (inconsistent) answer was actually given by the intended targets of the allegation. In the early calculus the inconsistency is even more obvious than in the case of Aristotle's theory, strongly suggesting that nobody would really have committed in the way Berkeley suggests. However, unlike the case of Aristotle's theory, there is no obvious way to avoid Berkeley's charge and still account for the phenomenon in question. In Aristotle's case there are numerous ways to proceed in a consistent fashion, both maintaining Aristotle's core belief about how weight affects the rate at which bodies fall, and rejecting this belief as Galileo did. But in the case of the early calculus nobody, including Berkeley, knew how to give a consistent and satisfying answer to Berkeley's question until one hundred years later (although many thought they knew, and one or two came close). However, this doesn't change the fact that in both cases the charge of inconsistency is not very fair. Galileo should have said that the Aristotelian's account of how combined bodies fall was not conceptually satisfying (and didn't match what is observed about such bodies). Berkeley should have said that no successful justification for the moves made within the early calculus had been given, not that people had an inconsistent justification. Both Galileo and Berkeley exaggerate for rhetorical purposes (see §8.2.2, this volume).

The questions in the cases of Olbers' paradox and classical electron models have a similar function to those seen in the cases of Aristotle's theory and the early calculus. In all four of these cases a question is asked which *could* lead to a commitment to an inconsistent set of propositions, but needn't do so if one is moderately careful. In the early calculus the propositions weren't committed to (except possibly by Johann Bernoulli) because the inconsistency is immediately obvious. In Aristotle's theory and classical theories of the electron the inconsistency is similarly very obvious. Things are a bit different in the case of Olbers' paradox, though. This time there is such a multitude of ways in which it might turn out that the night sky *should* be dark after all (e.g. because there are gaps in the ether) that inconsistent commitment is all but impossible. Putting together a set of properly inconsistent propositions in this latter case would be to indulge in a radically a-historical rational reconstruction of science. Unfortunately, despite Kuhn and the 'historical turn' in the philosophy of science, such a-historical 'rational reconstructions' are still all too common, as we saw for example with paraconsistent reconstructions of the early calculus.

Finally a note on CED. This case differs from the other seven, because there is no obvious question which either brings the inconsistent propositions together (as in Newtonian cosmology, the early calculus, Aristotle's theory, Olbers' paradox, the classical theory of the electron, and Kirchhoff's theory), or which leads to the derivation of a contradiction from propositions which are already 'together' (as in the later Bohr theory, or Poincaré's question about the predictions of Kirchhoff's formula at the aperture and behind the screen). Instead the catalyst for bringing the inconsistent propositions together is twofold: (i) the character of the domain of phenomena in question, and (ii) the pragmatic usefulness of the relevant propositions to solve problems of a particular type. We saw in Chapter 4 that there is a very natural domain of EM phenomena which demands explanation, but which includes self-*fields* whilst excluding self-*forces*. It turns out that the latter, but not the former, are (almost always) insignificant in the domain. This encourages one to work with a set of propositions which, if taken at face value, are inconsistent. But more importantly, even for problems where self-force effects are *not* insignificant, the obviously-false Frisch-LFE is still used. So it is not always the case that an inconsistency comes to light when one asks a particular question: just as propositions can be drawn together to answer a question, propositions can also be drawn together to provide modelling tools for a domain of phenomena.

Thus, if it were deemed worthwhile to systematically search for inconsistencies in science, it would be equally worth looking for new questions to ask, and new ways to specify domains of phenomena to explain. But it should be kept in mind that just as a question can be based on a false premise, a specified domain of phenomena may be artificial. For example, Shapere (1984b) considers 'Problems of Domain Coherence',

[h]aving to do with whether the domain does indeed constitute a domain: with whether, for example, a unified account of all the items so grouped really is to be expected, or with whether some particular item really belongs in the domain.... The possibility of such problems arising shows clearly that the claim that a certain body of alleged information constitutes a domain is a hypothesis, subject to rejection in the light of new discoveries. (Shapere 1984b: 642)

So it may turn out that although a given set of inconsistent propositions *appear* to be 'pointedly grouped' in a significant way, they are not because the domain that unites them is, to use Shapere's word 'superficial'. In Chapter 4 we saw that it is not obvious where to draw the boundary of the domain of EM phenomena that CED should explain. For example, one might have reason to exclude phenomena where self-force effects are significant, especially since no such phenomena were observed until the first synchrotron accelerator was built.[1]

'Superficial' domains might be compared with questions based on a false premise. The most obvious candidate for that in the eight case studies is the question asked in Newtonian cosmology, which assumes that the *net force* on an object is something that must exist. In this case, the false premise on which the question was based was actually an assumption of the theory. This, again, is a feature unique to only one of the case studies, and further warns against substantial generalizations about inconsistency in science.

Suppose we were to embark on a quest to find the next important inconsistency in science. Well, many of the case studies have not turned out to be especially important inconsistencies, so it is little wonder that they are little use on this point. Perhaps the best we can do is list the different ways in which inconsistency has 'come to light' in the case studies, and hazard a guess that the next inconsistency may emerge in one of these ways. It may be via a question which carries on work in a domain which is already being investigated, such as in the later Bohr theory. Or it may be via a question which pushes our science into new domains, and forces us to bring together propositions usually isolated in their separate domains. This would be like Seeliger's question, which brought together assumptions from

---

[1] Clearly much more could be said about the nature of domains, but since they have played a relatively small part in the cases studies I leave this for another day. The interested reader should consult Shapere (1977, 1984b) and Nickles (1977).

cosmology and mechanics, or the question of the classical theory of electrons which brings physics into contact with issues usually coming under the bracket of 'metaphysics'. Alternatively it may be a question which imagines an entirely new scenario not yet considered, such as Kirchhoff's question of the behaviour of light in a particular experimental setup, or Galileo's question about the effect of gravity on 'combined bodies'. And what makes the job harder is the fact that sometimes the crucial question is not obviously an interesting one to ask, such as Pauli's question of Bohr's theory of the atom, question (Q') in Newtonian cosmology, and Kirchhoff's question insofar as it is a question about an impossible, idealized experiment. Further, it may not be a case of asking a specific question at all, but rather a case of considering a newly specified domain of phenomena.

Perhaps nobody ever expected us to find a set of rules for unearthing the next important inconsistency in science. But nevertheless, the extent of the differences between the cases is worth emphasizing. We have seen that there are a range of different types of question, leading to a range of inconsistencies with different characters, and which matter to a greater or a lesser extent depending on the material facts of the particular case. In fact, in many of the cases it is doubtful that there is any good sense in talking of an 'inconsistency in science' at all. This thought leads to one phenomenon which *does* seem to be quite common: the tendency to use the word 'inconsistent' to exaggerate the strength of one's claim.

8.2.2 *Inconsistency versus implausibility*

Some of the cases could sensibly be reinterpreted in terms of implausibility rather than inconsistency. For example, a critic of Bohr's theory could take the viewpoint that the central propositions were merely implausible, rather than inconsistent, because of how they jar with CED (the discreteness of the energy states, the mysterious quantum transitions, and the non-emission of radiation from a charged, orbiting particle). And one could take the view that Aristotle's theory of motion becomes implausible, rather than inconsistent, when one considers combined bodies (e.g. the falling bag of peas example), because it forces upon us fantastic propositions concerning the entification of bodies. Similarly with many of the other cases. One option would be to ask the question whether these 'theories' are *really* inconsistent or instead just implausible. There are good reasons, I suggest, for thinking that there is no fact of the matter regarding this question. This is because there is no fact of the matter as to whether all of the relevant propositions really do belong 'in the theory'. However, as I've already noted, I do not need to make this argument to show that framing our questions in terms of 'theories' is a misguided approach. The debate is much more transparent if we eliminate *theory* on pragmatic, and not necessarily metaphysical, grounds (see §8.3, below, for more on this point). One can group together mutually inconsistent propositions, or other propositions which might reasonably be described as mutually 'implausible', if one wants to. The real question is what the point of grouping propositions in that way would be. What do we learn about how science works, or could work?

There is an important point to make here about rhetoric. Philosophers, scientists, and human beings in general want to make bold, important, influential claims. If one finds that a set of propositions are inconsistent then there is *definitely* something wrong, definitely an assumption which has to be rejected as false. If one instead claims that a set of propositions are merely 'implausible', the response might be that many scientific truths are initially regarded as implausible, such as that the surface of the earth is moving at many hundreds of miles per hour, that space can be 'curved', that spatially separated electrons can be 'entangled', and so on. Thus mere implausibility leaves things unclear: perhaps there really is nothing wrong with the assumptions. If one really wants to reject the assumptions outright, then inconsistency is preferable.

So consider again Aristotle's theory of motion. Galileo clearly wanted to falsify the theory, and so he would have wanted to make the stronger 'inconsistent' claim. This he in fact did explicitly (see for example Schrenk 2004). But Galileo also wants to make a strong claim in another way: he wants his claim to be about 'Aristotelian theory'. The problem comes if we eliminate *theory*, and ask the question of what the justification is for grouping together propositions in such a way that they really are inconsistent here. We saw in §7.2 that in order to do this one must add the assumption 'Natural speed is mediative'. But then the question arises whether any of the Aristotelians really made a commitment to this assumption, whether it has any historical relevance. This quickly leads to the thought that there really isn't much point in grouping the inconsistent propositions together to try to make any point about 'how science works'.

In short, the problem is that philosophers, scientists, etc. want to say that a claim is important because it is about a *theory*, and also that it is important because it is about *inconsistency* (as opposed to the more subjective 'implausibility'). Galileo succumbs to the temptation. But if we eliminate *theory*, and ask which propositions we really care about in this case, it is those that were actually, seriously committed to by Aristotelians, and this *doesn't* include 'Natural speed is mediative'. Thus it seems that Galileo wants to have his cake and eat it: he wants to say that he is attacking assumptions central to the Aristotelian's position, but he also wants to say that these assumptions are inconsistent. One of these claims must be weakened: *either* one can select propositions the Aristotelian definitely committed to and argue that they lead to an implausible conclusion, *or* one can add the negation of this implausible conclusion to the original assumptions, and then show that the resultant assumption set is inconsistent. In this thesis I have chosen to analyse the case in the latter way, since Aristotle's theory is widely labelled 'inconsistent' in the philosophy of science literature. What we then find, of course, is that the inconsistency is not very important or interesting from the point of view of history and philosophy of science.

The classical theory of the electron is a similar case in this respect. Landau and Lifshitz, Feynman, Shapere, Frisch, and others use the word 'inconsistent' here. And they also suggest, or state explicitly, that the inconsistent entity is 'classical (relativistic) electrodynamics'. But again, one of these two claims must be weakened. I argued in §7.4 that it would be quite odd to define CED in such a way that it takes a stance on whether electrons are elementary. But if it doesn't, then one should say not that CED is internally inconsistent, but that it is inconsistent *with* the claim that electrons are elementary. And then the conclusion is less dramatic, since in the early years of the twentieth century it was common to suppose that electrons have parts and are deformable, and so are not elementary. Olbers' paradox is similar, although in this case one needs to add *many* propositions to the central assumptions to reach inconsistency. In both of these cases even 'implausible' might be too strong, since *at the time* the extra propositions needed to reach inconsistency were not especially implausible.

There are one or two other examples of this propensity for scientists to overstate their case which haven't been considered at all so far. For example, during the Leibniz–Clarke correspondence Samuel Clarke wrote,

That one body should attract another without any intermediate means, is indeed...a contradiction: for 'tis supposing something to act where it is not. (Alexander 1956: 53)

But as Lange (2002) comments,

[T]hat a body acts where it is not (action at a distance) leads to a contradiction only when combined with the assumption that a body is present wherever it has an effect (spatial locality). Together these entail that a body is present at a location where it is not present. That is a contradiction! (Lange 2002: 95)

Clearly the target of Clarke's claim—Leibniz—would not have claimed that a body is present wherever it has an effect. So Clarke has overstated his case in the same way Galileo did: either he should say that an *implausible* conclusion follows from Leibniz's claim, or he should say that Leibniz's assumption is inconsistent with the negation of that implausible conclusion. But then this is something Leibniz would have agreed with (except, perhaps, for calling it 'implausible').

So this seems to be a fairly widespread phenomenon, and accordingly any claim that a 'theory' is/was inconsistent should be very carefully judged. And this lesson extends to cases typically referred to as *mutual* inconsistency, between different theories. For example, Priest (2002: 122) brings up the case of the 'conflict' between thermodynamics and evolutionary theory. The claim is that, according to evolutionary theory, the sun must have been providing heat and light for the earth for hundreds of millions of years, whereas according to thermodynamics this is impossible. However, to reach inconsistency one must include the assumption that the sun's heat is *not* generated in some as yet unknown way. But was it so implausible, in the nineteenth century, that the sun's heat was generated in some unknown way? It is far from clear that there is a good point to making the claim that this is a case of 'two mutually inconsistent theories in the history of science'.

This habit of overstating one's case is an essentially human thing, not exclusive to science and philosophy of science. A non-scientific example—particularly poignant given recent developments in the health of the economy—was emphasized by Rescher back in 1973:

[W]hile Marxists are constantly expatiating upon the 'contradictions' of capitalism, it is clear that in effect they mean no more than that capitalist societies exhibit deep internal conflicts, mutually destructive tendencies, and inner antagonisms.

Such Hegelian so-called 'contradictions' have to do with inconsistencies and instabilities, with inner tensions and strains and opposite pulls. There is nothing actually *inconsistent* about any of this in the logically rigoristic sense. Outright self-contradiction as the logicians explicate it—as transgressions of the Law of Contradiction—is not at issue. (Rescher 1973: 88)

Again there are really two options. Define 'capitalism' in a modest way which will be quite widely accepted, but which will mean that it is not inconsistent, or explicitly augment 'capitalism', including further propositions such that it *is* inconsistent. Neither set of propositions will be what capitalism *really* is—or at least, there is nothing to be gained by trying to argue that one particular set of propositions *is* 'capitalism'. Instead one can avoid arguments about the definition of capitalism, and just explain what it is we learn from the fact that this or that particular set of propositions is inconsistent. If the point was to try to show that many people ('capitalists') have an inconsistent belief set, then the reconstruction will not succeed: one will not be able to put together a set of propositions all of which are widely believed, and which are sensibly described as 'inconsistent'. This is to distort the facts to make an overblown, sensationalist claim for rhetorical purposes.

In this final example, as with the examples of scientific theories, the name we call our set of propositions does more harm than good. Miscommunications and arguments concerning what a given name refers to would be immediately eliminated if we eliminated *theory* and theory-names. One would then simply have to say what exactly is interesting or important about the fact that a given set of propositions are inconsistent, or implausible. This would make it very much more difficult to distort the facts and exaggerate one's claims.[2]

8.2.3 *The role of 'as if' reasoning*

In the case studies we saw several examples of 'as if' reasoning. Most obvious was the 'fictionalist' attitude to the infinitesimal in the early calculus of Leibniz and others (§6.5). But then also recall Debye's attitude to certain

---

[2] Kant referred to 'antinomies', where 'we find contradictions which arise when we attempt to apply certain concepts in a pervasive way, that is, to the full extent of their possible application', as Taylor (1977: 228) puts it. The claim might then be that certain claims seem implausible because when the relevant concepts (such as *capitalism*) are applied to their full extent one finds inconsistency. Of course this all hangs on whether there is something genuinely useful in the idea of 'applying concepts to their full extent'.

features of Bohr's theory of the atom: his view was that, although one reasons *as if* electron orbits enjoy a quantized spatial orientation, this is really to be interpreted as a mere instrument for calculation, a 'timetable for the electrons' (§3.3.2). Then in Newtonian cosmology the inconsistencies were eventually overcome when gravitational force came to be interpreted as a gauge quantity without direct physical significance (§5.3.2.3). In other words, although we still reason *as if* bodies are affected by a 'really existing' gravitational force, there is an anti-realist interpretation of this which eliminates the threat of inconsistency. Also one might consider here Kirchhoff's derivation of his diffraction formula, which treats the screen *as if* it is infinitely thin and infinitely opaque, when of course this is really considered to be an idealization of any real-life situation (§7.5.2). And finally one might consider as 'as if' reasoning any of the other idealizations which have been mentioned along the way. For example, in CED reasoning proceeds *as if* $\mathbf{F}_{self} = \mathbf{0}$, when in fact it is really $\mathbf{F}_{self} \approx \mathbf{0}$ which is believed (§4.5), and reasoning often proceeded *as if* electrons are points in the early twentieth century when scientists really believed that they are extended (§7.4.2).

There is an important relationship between 'as if' reasoning and inconsistency. In all of these cases the 'as if' commitment to the relevant propositions means that any inconsistency, or potential inconsistency, is not the serious problem it otherwise would be. In particular, in such cases there is no doxastic commitment—no commitment to the assumptions as true or even as candidates for the truth—and so no 'crisis' of the Pauli and Newtonian cosmology (I2) type. However, once again we find that the differences between the cases are striking. In certain cases we know where the truth lies, so to speak, but we depart from it for the sake of simplicity, to make calculations tractable. The CED (Chapter 4) and Kirchhoff cases are like this: in both cases the inconsistency is rendered innocuous because the derivation of any contradiction can be dismissed as an artefact of the idealizations. In both cases the de-idealization cannot be carried out due to computational intractability, but it is reasonable to assume that *if* the de-idealizations were possible, the inconsistencies would disappear.

The early calculus is a similar example in some respects. There were *some* in the community—Leibniz in particular—who considered any use of the 'infinitesimal' a shorthand for a more complex account which would refer only to finite quantities. So we have here a simplification for the sake of computational convenience akin to the CED and Kirchhoff cases. But on the other hand, this case is *very* different to those other two cases. The simplification is not a simple matter of turning an approximate equality into an equality, but rather the employment of a fictionalist metaphysics very different from the assumed reality. Then there were many different attitudes in the community, including a range of kinematic interpretations in terms of fluxions, fluents, moments, etc, a realist commitment to infinitesimals as variable quantities (in a sense), and even a realist commitment to infinitesimals as static quantities (Johann Bernoulli). And in all of these cases reasoning continued according to algorithms, so that the inconsistent propositions were never allowed to get purchase to reach contradictory conclusions. In this case, unlike CED and Kirchhoff, nobody (except perhaps, if we are feeling charitable, Newton and Leibniz) knew where the truth did in fact lie.

Despite the differences, in these three cases the 'as if' commitment means that a serious, problematic inconsistency never gets off the ground. But sometimes one can be faced with a serious, problematic inconsistency and *introduce* 'as if' commitment to eliminate the problem. This is the situation with Debye's attitude to Bohr's theory, and the modern attitude to Newtonian cosmology. Take the Pauli inconsistency of Bohr's theory. Adapting Debye's approach to the 'spatial orientation' of electron orbits, one might have said that electron orbits in general were not to be interpreted as 'real'. This would then mean that the derivation of an 'orbit' going straight through the nucleus would not be intolerable, since it wouldn't mean that an electron *really was* going straight through the nucleus. This in turn would mean that Sommerfeld's stipulation in his quantum condition $m \neq 0$ would not be necessary, and the problematic inconsistency would disappear. However, this would render the vast majority of the theory non-realist and instrumentalist, and most philosophers would see this as dramatically reducing its explanatory power. This would not be a way to doctor one's theory to eliminate inconsistency, but rather a way to change one's attitude to theorizing in general.

Of course, as we saw in Chapter 3, Debye never did suggest that the inconsistency be eliminated in this radical way. What he did suggest was that one or two propositions be interpreted in a non-physical way. This sort of piecemeal reinterpretation can sometimes be an extremely effective way to eliminate a problematic inconsistency, as the Newtonian cosmology case shows us. In that case the trick was to reinterpret the gravitational acceleration of bodies as a gauge quantity without direct physical significance—as Norton (1995) puts it, 'acceleration is relative'. Compare the modern attitude to movement: we now think it is a bad question to ask 'Is the sun *really* moving?', since the truth is that it is moving relative to some things, but not relative to others: there just is no 'really moving'. In modern Newtonian cosmology (Malament 1995) this is extended to the question of the acceleration of bodies. And this is enough to eliminate the inconsistencies of Newtonian cosmology presented in Chapter 5, whilst avoiding a wholesale 'instrumentalist' interpretation of the other key propositions of that chapter.

Perhaps, then, there is something to learn here for the resolution of current conflicts, such as that between general relativity and quantum theory. Perhaps there is an assumption the physical interpretation of which can be re-assessed. On the other hand, given that many of the supposed cases of 'inconsistency in science' considered here have turned out *not* to be cases of inconsistency in science in any serious sense, another look at the conflict between general relativity and quantum theory may be in order.

8.2.4 *Getting from inconsistency to contradiction*

Some of the case studies show us that from time to time there is an implicit inconsistency in our science, in one form or another, but that it is not appreciated for many years. In the case of Bohr's later theory there was an implicit inconsistency for eight years, with Kirchhoff's theory there was an inconsistency for ten years, and in the case of Newtonian cosmology one might say that there was an implicit inconsistency for two hundred years. In cases such as these we want to know what prevented the inconsistency from being appreciated sooner, and so what might be preventing us from noticing an implicit inconsistency in our *current* theories.

Of course sometimes an inconsistency is not appreciated because the relevant propositions are not put together. This might depend on our asking the right question, as noted in §8.2.1. Another possibility is that one of the relevant propositions is implicitly committed to in the sense that it is guiding scientific reasoning but scientists are not explicitly aware of it. For example, in §3.2.1 I mentioned the possibility of a 'continuity principle' at play in classical physics at the beginning of the twentieth century, and in §7.4 I mentioned the possibility that scientists were implicitly committed to the impossibility of an infinite and negative 'bare' mass of electrons. In these two cases scientists had no trouble noticing the inconsistency in question, but this is because the implicit propositions were guiding their thoughts, not because scientists were explicitly aware of them. This given, the scientists were also not aware of one possible way to avoid inconsistency: rejection, or modification, of the implicit proposition in question. In the case of classical electron models it wasn't until Dirac in 1938 that the implicit proposition in question was isolated and rejected as a way to get around the relevant inconsistency: perhaps electrons *could* have an infinite and negative 'bare' mass after all.

With the case of classical electron models in mind, one may wonder how we might speed up the process of isolating implicit propositions which are guiding science, and which we might want to reject as science throws up conceptual difficulties or leads us into contradiction. Drawing on the case of classical electron models, one option would be to think more carefully about why a given route of enquiry is being dismissed. What initially seems absurd (point electrons) might end up being more plausible than the alternative. Lorentz, for example, eventually gave up on a realistic interpretation of his extended, deformable electrons, even with a Poincaré counter-pressure in place. Perhaps more useful here than the advice 'think more carefully' is for scientists to sometimes try to reconstruct their reasoning in the philosopher's way, as a formal deductive argument from a set of propositions. This is often an effective way to identify the background assumptions one is making, which turn informal reasoning into something the logician could accept as 'valid'.

But the main question for this section is why scientists sometimes fail to notice an inconsistency, even after the propositions have been 'put together' more or less explicitly. Of course an inconsistency is noticed when and only when one sees that a contradiction follows. Three main things get in the way of our noticing that a contradiction follows from a given set of propositions:

(i) The derivation is long and/or complicated.
(ii) The derivation depends on inferences we are not familiar with, or which we don't recognize as truth-preserving.

(iii) The derivation is motivated only when we ask a very specific question.

The Pauli inconsistency combines (i) and (iii) here: although the propositions in question had *already* been put together as relevant to the domain of atomic phenomena, they hadn't been combined in the right way to reach contradiction. But even then, a long and complicated derivation was required; no wonder that it took eight years for the inconsistency to be noticed. The lessons from this particular case for current science can only be to search for new questions to ask, and to persevere with long and complicated inference chains.

Kirchhoff's case is slightly different. Noticing the inconsistency does not depend on a particularly difficult derivation. However, it does depend on asking a very specific question: what does Kirchhoff's diffraction formula tell us the intensity of light will be at the aperture and immediately behind the screen? In Kirchhoff's time one would have been more interested in investigating the successes of Kirchhoff's formula than asking apparently useless questions like this. For one thing, the intensity of light at the aperture and behind the screen is already stated in the assumptions of the theory, so one would have thought that there was no need to see what the derived diffraction formula says about the intensity there. In addition, Kirchhoff derived the diffraction formula to find out what the intensity of light would be *away* from the screen—this is what scientists were really interested in. In addition, we may ask *which* screen the question pertains to. Is it the idealized, infinitely thin screen Kirchhoff imagined, or some real-life screen? Why we should care about the former isn't clear, but if the question is about the latter the inconsistency does not apply, because the propositions from which the contradiction can be derived are only relevant to the idealized screen. So it is no wonder it took ten years for Poincaré's question to be asked.

A more interesting case is the case of Newtonian cosmology. Here the derivation of a contradiction is relatively straightforward as seen in Chapter 5, and in addition some relevant individuals (such as Newton) *did* ask the question which leads to contradiction. What prevented the discovery of inconsistency in this case was the difficulty of recognizing as truth-preserving what I labelled inference (I★):

(I★) from indeterminacy infer that there is no solution.[3]

This happened mainly because the mathematics involved was not yet properly understood. Neither would it be properly understood for a long time to come, at least when Newton was writing. But there is a separate issue which should also be recognized. In this particular case it is extremely easy to conflate two different meanings of 'no force'. One is that the magnitude of the force is zero; the other is that *force* doesn't exist, in a metaphysical sense. It is only when one recognizes that the latter follows from the relevant propositions that one reaches inconsistency. So it was extremely easy to overlook the particular inference involved, even after the relevant mathematics had advanced in the nineteenth century (as we saw in §5.4.2).

The lesson in this case? To be careful to note that in some mathematical scenarios it can be very difficult to see what should count as 'truth-preserving', and also how we should interpret the mathematics physically. Consider the following case as a different example, where '$\sqrt{}$' denotes taking the *positive* square-root (from Wilson 2006: 312):

$$2=\sqrt{4}=\sqrt{-2.-2}=\sqrt{-2}.\sqrt{-2}=\sqrt{-1}.\sqrt{2}.\sqrt{-1}.\sqrt{2}=(\sqrt{-1})^2.(\sqrt{2})^2=-2$$

If one follows this line of reasoning it seems that every step is truth-preserving. But if that were the case it would be true that $2=-2$! Where have we gone wrong? Is it that '$\sqrt{}$' is not distributive after all? Or is it something to do with the fact that the proof postulates square-roots of negative numbers? Little doubt that such cases will continue to infiltrate the corners of mathematics. As mathematics expands into weird and wonderful new areas, far beyond the reaches of our intuition, the very idea of 'truth-preservation' may be in doubt. Indeed, what we mean by 'truth' in mathematics is an on-going area of debate (see e.g. Dales and Oliveri 1998). For the 'formalist' there is no truth in mathematics, but instead uninterpreted axioms, rules of inference, and theorems. For some, the motivation for picking one set of axioms rather than another is based, in part at least, on its scientific usefulness. This leaves open the possibility that one could turn an inconsistent scientific theory into a consistent one by adjusting the axiomatic

---

[3] Or no solution in any physically relevant type of number, at least.

basis of the mathematics in play, on the grounds that it is not useful to derive a contradiction!

A slightly different complication is that, in addition to deciding what should count as truth-preservation in mathematics, there is also the question of how to interpret the transition from the physics to the mathematics in the first place. Consider, for example, Newton's apparent claim that there aren't really unidirectional gravitational forces in nature, but only inseparably mutual gravitational interactions *between* bodies (recall footnote 28 from §5.4.2). If this were right, then unidirectional vector representations of such 'forces' would only be useful mathematical tools, not to be taken as physically literal. One might then attempt to justify ignoring a peculiar consequence—such as the derivation of an indeterminate net force on the Earth—on the grounds that it is an artefact of using a mathematics which departs in certain ways from the true physical situation. If one takes this attitude to the relationship between physics and mathematics in general, then one might convince oneself that any derivation of a contradiction from a scientific theory can be blamed on one's mathematics instead of one's physics. It is well-known that physicists often derive two solutions to a problem, but then reject one of them as 'unphysical', perhaps because of the presence of a complex number in the solution. In such circumstances, if one believes one's theory to be a candidate for the truth, and one believes that all inferences have been truth-preserving, then the only justification for ignoring the solution can be that the mathematics in play is an imperfect representation of the original, material, non-mathematical assumptions.

My intention here is not to explore these issues in the philosophy of mathematics in detail,[4] but just to emphasize why philosophy of mathematics matters to the question of inconsistency in science. In short, whether a given theory is a contender for the truth of our world may depend on the advancement of philosophical disputes about truth and truth-preservation in mathematics. Thus the importance of such debates is clear: we do not want to end up with another case like Newtonian cosmology where a misunderstanding about mathematics (in that case how we should think about infinite, non-converging sums) means that we overlook an absurdity in our

---

[4] The interested reader could do worse than to start with Sarukkai (2005), as well as the Dales and Oliveri (1998) volume already mentioned.

scientific hypotheses (our beliefs, even) for many decades. To make this more than merely hypothetical, one need only turn to quantum field theory, where the relationship between the physics and the mathematics is still poorly understood. Most significantly, to get the superbly accurate predictions out of the theory it is apparently necessary to use non-rigorous 'informal' mathematical techniques (e.g. renormalization techniques). No mathematically rigorous, physically realistic formulation of quantum field theory is currently known. Whether this means there is a problem with the *physics* is an area of open debate.[5]

8.2.5 *The logic-driven/content-driven debate*

How do the case studies bear on the logic-driven/content-driven debate introduced in Chapter 1? Recall that the point of that debate was to decide whether, when faced with inconsistency, scientists either *do* or *should* restrict their reasoning by changing their logic, or by restricting their inferences in some other way depending on the precise nature of the content involved.

First of all, we have seen that the number of significant inconsistencies which exist in real science is far less than has often been suggested. The main examples of significant inconsistencies in this thesis are the later Bohr theory of the atom, Newtonian cosmology, and (to a lesser extent) the early calculus and Kirchhoff's theory of diffraction. But in each of these cases it is clear that scientists don't avoid the conflict by changing their *logic*. As the case studies have shown, classical deductive logic is used. Nor does it seem plausible that scientists *should* handle these cases by changing their logic. All that might be said is that these cases *can* be reconstructed by changing the logic, but it is hard to see what understanding might be gained by such a reconstruction. We saw in Chapter 6 that some paraconsistent logicians have tried to 'rationally reconstruct' the early calculus. Brown and Priest (2004) even state that they are not trying to be historically accurate, but just trying to 'capture the practice' of mathematicians at that time. But as I argued in §6.6, it isn't at all clear how the paraconsistent reconstructions really do 'capture the practice' in some interesting or important way.

---

[5] See Fraser (2009) for a recent perspective on these issues in the foundations of quantum field theory.

So in other words this thesis supports the case of those who advocate a content-driven approach. However, it only supports it so far. For example, it doesn't support the idea that, when explicitly faced with an inconsistency, scientists either do or should restrict their inferences according to the content. For example, we see that in the case of the later Bohr theory as soon as the inconsistency was made apparent scientists dropped an assumption, as Lindsay's (1927) paper attests. Dropping an assumption is obviously something quite different from keeping the same propositions but restricting one's inferences to avoid absurd results! Similarly in the case of Newtonian cosmology, after Seeliger made the inconsistency apparent in 1895 there were a multitude of attempts to change one or another assumption (see Norton 1999). How the scientists avoided contradicting themselves in these two cases is clear enough. In the case of the later Bohr theory one simply doesn't reach inconsistency unless one makes a particular and detailed derivation. In the case of Newtonian cosmology individuals didn't realize that they could make inference (I*). This latter case could perhaps count as a case of inferential restriction, but it would be a case of *unintended* inferential restriction. Davey (2003: fn.16) cites Newtonian cosmology as an example where physicists use an 'inferentially restrictive methodology'. However, Davey fails to note that this case certainly doesn't support his claim that physicists *knowingly* use an inferentially restrictive method, and that they *should* use an inferentially restrictive method. In this case they didn't know their inferences were restrictive, and it would have been a big step forward had this been exposed and acted upon.

Kirchhoff's theory is different: in this case it really does look like scientists continued to use the theory and tolerate the inconsistency. The inconsistency is avoided simply by stipulating that the diffraction formula doesn't apply at the aperture and behind the screen. In other words, a domain of application is specified—determined empirically—where one can expect sensible results. As discussed in §7.5.2, the restriction on the domain of application is not ad hoc in the sense that it arbitrarily covers up an underlying problem with the theory. It is known from the beginning that the theory is idealized, so that what is stated by the final formula isn't going to be a candidate for the truth of any real system. How then are inferences restricted? Since the theory can't be de-idealized it can't be known which predictions to believe and which not to believe in the final formula, except that one knows to ignore contradictions, or especially wild predictions.

Other than this, it seems best not to talk in terms of 'inferential restrictions' at all, but rather to simply say that where the diffraction formula has been tested and seen to closely match the world, it is sensible to trust it, and where it is known to fail (e.g. near the screen), it shouldn't be trusted. As unsatisfactory as this is, since the theory can't be de-idealized it seems to be the only sensible way to restrict one's inferences. Whether this counts as content-driven control depends on how you want to define 'content-driven control' (recall §1.2.1), but we can at least say that there isn't any *logic*-driven control going on here. In the end it might be more sensible to say that there is nothing that really *needs* 'controlling'.

Finally we may look to the cases of CED (Chapter 4) and the early Bohr theory considered in §3.2.3, both of which have a similar relationship with scientific idealization to Kirchhoff's theory. In all of these cases scientists didn't commit to all of the relevant propositions as candidates for the truth, but they nevertheless reasoned with the propositions *as if* they committed to them as candidates for the truth (as discussed in §8.2.3). As in Kirchhoff's case, in Bohr's theory and CED it is clear that scientists did not change their logic, nor ought they to have changed their logic. The scientists in question knew that they were working with (at least some) false propositions, and so judged all of their inferences carefully. Any contradictions could be put down as artefacts of the idealizations in play. How to move forward in such circumstances? With difficulty, especially when full de-idealization is not a serious possibility. CED is an interesting case here, since very serious attempts at de-idealization have been made over a period of several decades, with partial success (see Frisch 2005a: ch.3; Muller 2007: §§6 and 7; and Frisch 2008). Partial de-idealization might increase one's confidence in any conclusions drawn from truth-preserving inferences, but the basic situation remains the same: scientists still proceed carefully in full knowledge that they are working with propositions at least some of which are definitely false. Whether this sort of de-idealization is possible, and the precise way(s) in which it should be done, depends on the material facts of the case at hand. This suggests 'content-driven control' since the material content of the propositions dictates whether and in what way de-idealizations are attempted. But actually the phrase is still misleading since there isn't a real sense in which one needs to 'control' ones inferences to avoid logical explosion. Logical explosion does not seriously threaten in such cases at all, since scientists are nothing like blind deductive machines, particularly when they know for sure that at least one of the assumptions they are making is false.

This is brought out especially vividly in the 'toy' Goldbach case I discussed in §3.2.3.2. How is logical explosion avoided? Very easily. The real question is, how on earth could explosion be *achieved* in this case (assuming for the sake of argument that Goldbach's conjecture is false)? Or one can consider a similar real-life example: that of Kirchhoff's assumptions plus Maxwell's equations. These are inconsistent, but nobody knew it for the best part of 100 years. The real challenge is to get from inconsistency to contradiction here. The question of how reasoning is possible, and how one can trust any of one's inferences, is easily answered. One can reason as normal, and for the most part using truth-preserving inferences with approximately true assumptions will lead to approximately true (or even true!) conclusions.

Such considerations bear on Bueno's (2006) arguments for the use of paraconsistent logic in cases where we have a set of inconsistent propositions none of which we want to eject. In such cases one can actually keep all of the propositions, and continue to reason with classical logic, *and* still trust many or most of one's inferences. As just noted, the distance between inconsistency and explosion can often be considerable. In the meantime, using truth-preserving inferences to reason from approximately true assumptions—even ones which are mutually inconsistent—is a perfectly sensible way to try to stay as close to the truth as possible. And in many cases it is perfectly sensible to drop an assumption when one notices that one has reached a contradiction, as Lindsay suggests in the case of the later Bohr theory, or as several individuals suggested in the case of Newtonian cosmology (Norton 1999). In all of the cases I have considered, there is no case that fits with Bueno's suggestion that sometimes one will want to reason from an inconsistent theory with a paraconsistent logic, on the grounds that (i) one cannot sensibly use classical logic, and (ii) one doesn't know which assumption to eject. Thus it starts to look like this situation does not turn up in the real history of science, and is instead a philosopher's reconstruction of a hypothetical science that does not exist.

So in sum I suggest that the case studies here favour the content-driven approach over the logic-driven approach. But for most purposes neither approach is required because scientists do not work with, or do not make a serious doxastic commitment to, inconsistent assumption sets. Alternatively, neither approach is required when the inconsistent propositions in play are all (individually) either true or approximately true and thus lead to many sensible and reliable conclusions, and the lurking contradiction which follows from the propositions is *extremely* difficult to infer and exploit (even if we wanted to exploit it). But at least the content-driven approach acknowledges the diversity of science, stating that our handling of an inconsistency will depend on material facts of the case at hand which will vary from case to case. The logic-driven approach is often put forward as an approach that can 'capture' how inconsistencies in science are handled in 'theories' generally, or at least in many cases in the history of science. Brown and Priest (2004), for example, wish to advocate 'chunk and permeate' reasoning in several different cases of (alleged) inconsistency in science. But the details of how science works in practice, and the details of the history of science, do not allow for this sort of generalization. Theory eliminativism will help philosophers avoid making this sort of mistake.

8.3 Theory Eliminativism and the Method of Philosophy of Science

From time to time philosophers have reflected on the relationship between philosophy of science, science, and the history of science. Famously, Kuhn (1962) suggested that by not paying enough attention to the real history of science, philosophers of science were completely overlooking fascinating features of how science *really* works.[6] Before Kuhn there was a disconnect between philosophy and the history of science which was truly remarkable. In the last fifty years this has certainly been remedied to some extent, and many now declare that 'integrated history and philosophy of science', or 'iHPS', is alive and well (cf. Howard 2011). But even if iHPS is practised by *some* philosophers, it remains the case that much philosophy of science still doesn't pay much attention to the history of science, but remains in a bubble of purely philosophical literature. And there is another related but distinct problem: the disconnect between philosophy of science and *science itself*. That is, philosophers often fail to appreciate the true subtleties of how science

---

[6] By 'real history of science' I mean history of science at the 'pure' end of the reconstruction spectrum (recall footnote 11 in Chapter 2).

works (quite apart from its history). This is something Mark Wilson has done a great deal to emphasize (see Wilson 2006: 127ff and 178ff.; 2013). Classical mechanics, for example, is still today assumed to be a simple, axiomatic system that philosophers of science can safely describe as 'Theory T', a 'theory' which describes a certain class of 'possible worlds'. Philosophers often have in mind Newton's three laws when they invoke 'classical mechanics'. As Wilson (2009, 2013) shows, the thing we like to call 'classical mechanics' is a far more complex beast than this simple picture does justice to. Or, if one really *does* just mean to refer to Newton's three laws with the term 'classical mechanics', then one is referring to something that really can't do any significant scientific work on its own, and doesn't define a class of 'possible worlds' that is remotely interesting (at least from a scientific point of view).

Ever since Kuhn (1962) philosophers have, from time to time, urged us to engage more seriously with real history of science and (less often) with science itself. The temptation, even today, is still to do philosophy of science with rather little experience of serious science or history of science. However, urging philosophers to 'engage more seriously' with science and the history of science is rather vague. It doesn't tell us much at all about what exactly philosophers of science should do, or how they should do it. And no doubt some philosophy of science is perfectly good philosophy but *doesn't* require serious engagement with science and the history of science. Thus vague sound-bites such as 'Philosophy of science without history of science is empty' (Hanson 1962; Lakatos 1970b) are not helpful—the details matter. I am hardly the first to make this point: Larry Laudan and others found such 'sloganeering' tiresome back in the 1980s (cf. Laudan et al. 1986; and references in Schickore 2011).

Here theory eliminativism has significant virtues, although they are indirect and not immediately obvious. Recall from Chapter 2 that the original motivation for theory eliminativism was to prevent any possible disagreements about 'theories' from compromising other debates, and in particular debates about inconsistency in science. If theories are not mentioned then any disagreements about the definition, nature, structure, ontology, and content of 'theories' cannot affect the debate.[7] This still

---

[7] In Vickers (forthcoming) I justify this claim in some detail. The short version is as follows. Such disagreements will not affect the newly framed debate because one does not just eliminate the *word*

stands, but an equally important benefit of theory eliminativism comes from the way debates must be reformulated if we eliminate *theory*, and the new questions which must then be explicitly addressed. Before reformulation one talks in terms of 'theory', and there is an implicit assumption that what one is saying is important simply because it is *about a theory*.[8] Suppose we get rid of *theory*. Then we must state explicitly which propositions (or models, or algorithms, or whatever—one time for all) we are investigating and commenting on. But then, because these have been described only as a set of propositions, and not a *theory*, we cannot possibly expect anybody to take interest in our claims until we've explained precisely *why it is especially important to focus on those propositions*.

What will be the different answers to this important why-question? Whatever the case, it will become much more explicit why the claim is supposed to be interesting or important from the point of view of philosophy of science. It may be that it is appropriate to focus on some particular set of propositions because they have particular explanatory or predictive virtues, or it may be because there is a specific result one can derive from them, or some relationship that exists between those propositions and another set of propositions. These would all be purely scientific reasons for focusing on a particular set of propositions. But then there are also social or historical reasons why it might be appropriate to investigate and comment upon a specific set of propositions. Perhaps it is because individuals all believed those propositions, or 'accepted' those propositions (as empirically adequate, say), or all used those propositions in one way or another. In practice (and as we have seen in the case studies) the set of propositions appropriate to study will change in subtle or not-so-subtle ways depending on which of these options is chosen. Crucially, one will have to justify why the particular chosen set of propositions is an appropriate one to focus on *for the philosophical purposes in question*.

'theory', keeping the corresponding concept in the background guiding one's analysis. Instead the debate continues entirely without the concept *theory* and (crucially) without other concepts dependent upon the concept *theory*. The other concepts used do not depend on one's theory-concept, since they are more basic.

[8] Cf. Wilson (2006: 127): 'In rendering "theories" into schematic T's and T's, our syndrome puffs the humble word "theory" into something quite grand, without it being exactly clear in what its grandeur consists.... Of course, we enjoy patting ourselves on the back by claiming, when we have an interesting suggestion to offer, that we have laid down a "theory" in some grand, if amorphous, sense, for the term carries a more impressive ring than "intriguing proposal".'

For example, in the case of inconsistency in science we may ask just which sorts of sets of propositions it is going to be interesting to focus on, such that inconsistency will be a truly important result. For example, it will be interesting if we find inconsistency in a set of propositions which were *believed to be true* (or believed to be important candidates for the truth) for a significant period of time by a significant scientific community. What is interesting is the question of why it took so long to notice the inconsistency, and whether we might be missing inconsistencies in our current science in a similar way. The only case we've seen that was really like this was the Pauli inconsistency in the later Bohr theory, although the Newtonian cosmology case is closely related.

Another way in which an inconsistency might seem to be interesting or important is if we find it in a set of propositions regularly *used* by scientists for explanatory and predictive purposes. We might wonder how, in such circumstances, it can be rational to make inferences and trust one's conclusions. However, I have argued that this phenomenon isn't really as interesting or important as is often made out. For example, we find this happening in the case of Kirchhoff's theory of diffraction, but there is no epistemic difficulty in supposing the inconsistency to be an artefact of the idealizations in play. One assumes that one's assumptions are all at least approximately true, so that even if they are inconsistent one should still expect many derivations therefrom to lead to approximately true conclusions. Indeed, in the case of the hypothetical 'Goldbach' case that I discussed in §3.2.3.2, although one is reasoning with a set of inconsistent propositions, the vast majority of the deductions from those propositions will lead to *true* conclusions. Also in that section I argued that this sort of thinking justifies reasoning from both Bohr's postulates and CED, despite the conflict there.

The main point for present purposes is that, in eliminating *theory* and thus shifting the discussion to sets of propositions, one is forced to think harder about the scientific and historical status and properties of a chosen set of propositions which would make one's claims significant for our understanding of science. In other words—instead of the vagueness one finds in the clarion call 'pay more attention to science and the history of science!'—theory eliminativism forces upon us a process of debate-reformulation which includes a very specific way in which one's philosophical claims become more scientifically and historically engaged. And this is in addition to the other important benefit already mentioned: that post-eliminativism any disagreements and miscommunications about 'theories' can no longer possibly compromise the debate.

Of course, in this book the focus has been on inconsistency in science. To that extent its scope is rather narrow. But it also acts as a test case for the current status of the relationship between philosophy of science, history of science, and science itself. On this question the focus on inconsistency in science is as good a focus as any other. The overarching question is whether the disconnect emphasized by Kuhn back in 1962 still lingers in some way, whether philosophers of science in general are making suitably scientifically and historically informed claims. No doubt things are better than they were, but I am inclined to agree with Mark Wilson (2006: 178) that our discipline has not yet fully recovered from the days before Kuhn. In the context of debates about inconsistency in science there are several symptoms of this. Some philosophers still think it is sensible to search for something that could be called a 'theory of inconsistent theories'. Some propose that one or another paraconsistent logic can be a useful way to reason from 'inconsistent scientific theories'. This assumes that throughout the history of science there really are all these 'theories' which are sufficiently similar, in the relevant respects, such that one can say something general about how we should react to inconsistencies within them. But worse than this, most of the 'inconsistent theories' commonly put forward are not, on inspection, inconsistent in any significant sense after all. And where we really do find inconsistencies it is often clear, when one looks to the details of the science and the history of science involved, that a paraconsistent logic would not be helpful. The attitude of Brown and Priest (2004) is typical in this literature: *don't look to the history of science and expect to find anything like this there, but do take our philosophy of science seriously as 'capturing' the practice of scientists in some significant sense*. But this 'significant sense', this 'capturing', remains mysterious. And if one can't see the relevance of the analysis when one looks carefully at the history, any such 'significant sense' seems unlikely to exist.

But this isn't to say that it is only paraconsistent logicians who fail to engage sufficiently with science and the history of science. Most philosophers' activities are premised on the assumption that there is a certain uniformity to the history of science, so that one can talk about numerous 'theories' which are essentially the same kind of thing, and about which we can say something perfectly general. Indeed, making substantial, very general claims about the structure and function of scientific theories is in many philosophers' minds *what philosophers of science do*. For example, it is assumed that we can say that any theory that has achieved novel predictive success has turned out to be approximately true (by present lights), or that theories explain in this or that way, or are reducible one to the other in this or that way, and so on.[9] In the case of inconsistency in science, one might claim that one should control 'logical anarchy' by restricting one's inferences in a certain way which depends on the material content at hand. This allows for a certain amount of variation in scientific theories, of course: it assumes that how we restrict our inferences will depend upon the material differences between different theories. But it still assumes that we *will* have to restrict our inferences when faced with an inconsistent theory. In most of the cases I have considered, when one looks to the details of the science and/or the history, this just isn't true.

Most alarming of all is the fact that, in all this literature, a number of 'inconsistent scientific theories' turn up again and again as examples, when in fact, as we have seen, many of these examples are not inconsistencies in any significant sense. In many cases one reaches inconsistency only by grouping together a number of propositions at least some of which are not at all historically relevant. For example, if nobody ever thought that 'natural speed is mediative', who cares whether the assumptions Galileo criticizes are inconsistent? And if nobody ever believed in both Bohr's early theory *and* CED, at the same time, then why is Bohr's theory such an accepted example of an 'inconsistent theory'? As we have already seen, although there are *some* interesting similarities between some of these cases, the differences are substantial and important. And these differences are nicely brought to the surface if we eliminate *theory* and investigate sets of propositions. In this way the true diversity and disunity of science is revealed.

Philosophers of science can still do extremely important *local* work—that is, work on individual propositions or collections of propositions. Again, I find myself agreeing with Wilson (2006): the 'craving for generality' (Pincock 2010: 121) characteristic of most philosophers of science may well be to crave what is not there to be had. Norton (2011) makes some similar claims about philosophical theories of scientific confirmation: in his

---

[9] Recall the Suppe (1977) quote from Chapter 1: 'It is only a slight exaggeration to claim that a philosophy of science is little more than an analysis of theories and their roles in the scientific enterprise.'

view inductive inferences are licensed by 'local' matters of fact. However, this isn't to say that *all* philosophical work must be 'local'. The point is that the most substantial claims about 'how science works' will have to be local. This still leaves room for moderately substantial claims which are moderately general, and not-very-substantial claims which are very general. By eliminating *theory* I submit that negotiating this balancing act between the substantiality and generality of one's claims becomes easier to do.

With these benefits of theory eliminativism revealed there seems to be no good reason why it wouldn't also be fruitfully applicable in other quarters of the philosophy of science. After all, one will be brought to engage more seriously with science and the history of science *whatever* one is debating: there is no special role played by inconsistency here. And, further, disagreements about *theory* will no longer compromise debates after theory eliminativism: again, there is nothing special about inconsistency which ensures this. And there is no good reason to suppose that the kind of difference of opinion about 'theories' and 'theory-content' that we saw in the Frisch–Belot–Muller debate and in the Bohr case will not occur elsewhere in philosophy of science. These were disagreements about the content and function of theories, and initially had nothing to do with inconsistency: if these disagreements exist here, then one should expect such disagreements to exist throughout philosophy of science. The only question then is whether such disagreements really *matter* in other debates in the philosophy of science. Even if they don't, theory eliminativism may be worthwhile as a good way to ensure engagement with science and history of science. But I'm sure they do matter, in at least some other debates: indeed it would be startling if disagreements about *theory* did *not* affect other debates in the philosophy of science when so many debates put so much weight on the concept of a theory.[10]

Consider as one example the longstanding debate over the (in)determinism of 'classical mechanics'. Several authors have urged that it is indeterministic (e.g. Earman 1986; Hutchison 1993; Norton 2008), whereas others maintain that it is deterministic, such that one reaches the opposite conclusion only by misunderstanding 'what the theory is' (e.g. Arnold 1977; Korolev 2007a, 2007b; Zinkernagel 2010). And Wilson (2009) argues that

---

[10] Cf. Burian (1977: 30): 'The differences in their [philosophers'] reconstructions may indicate that they refer to different entities under the label "the Darwinian theory of evolution." Should this be so, any differences in their evaluations should be unsurprising, for one's evaluation of a theory depends on which entity one takes that theory to be.'

the theory is neither deterministic nor indeterministic, because there are different 'species' of classical mechanics, some of which are deterministic and some of which are not. Who is right in this case?

Suppose we eliminate *theory*, and also the theory-name 'classical mechanics'. Then one would have to present a set of propositions, show that they implied determinism/indeterminism, and then, crucially, explain why it is especially interesting or important that that *particular* set of propositions imply determinism/indeterminism. The only way to do this will be to engage in a serious way with the relevant science and/or history of science. For example, perhaps it's revealing because relevant individuals in the history of science believed those particular propositions to be true, but explicitly rejected indeterminism even though those particular propositions imply indeterminism. Whatever the case, two things have been gained by applying theory eliminativism here. Disagreements about *theory*—and in particular about what classical mechanics *is*—can no longer compromise the debate.[11] But also, one is forced to explain much more explicitly precisely why the determinism/indeterminism result is an interesting or important result. And it isn't clear to me that there are any real costs involved in applying theory eliminativism to this debate: philosophers are forced to work harder, but for very good reasons! One can hardly claim that in eliminating *theory* philosophers will end up saying exactly the same thing but in a more convoluted and less transparent way. Philosophers do end up saying something more wordy, but also something ultimately *more* transparent because the concepts involved are cleaner, simpler, less susceptible to multiple interpretations.[12]

It isn't my intention to here investigate the prospects for the application of theory eliminativism in the debate about the determinism of classical mechanics, nor in a range of other debates which could be mentioned. The point for now is just that there is good reason to believe that there are at least some (perhaps many) other debates out there which could benefit

---

[11] Cf. Wilson (2009: 174): '[M]uch contemporary commentary on philosophical theories of matter in the eighteenth and nineteenth centuries strikes me as greatly compromised by its inclination to assume that phrases such as "classical mechanics" or "the Newtonian picture" capture surgically precise meanings, when, in fact, such terminology can be readily applied to deeply incompatible doctrines.' Cf. Burian (1977: 37).

[12] Wilson (2006: 128) finds certain appeals to 'theory' in philosophy of science 'counterproductive and obscurantist', and provides two examples. With theory eliminativism such obscurantism disappears, of course.

from this strategy. And, crucially, other *types* of debate. It remains possible that theory eliminativism could benefit any debate in the philosophy of science which puts weight on the concept *scientific theory*. And this will be the majority of the debates in the discipline: think of the pessimistic induction on past 'theories', the underdeterminism of 'theories' by evidence, the confirmation of 'theories', the equivalence of 'theories', the reduction of 'theories', and inter-'theory' relations generally. In particular, if what I have said about the 'disunity' of science is even half-way right, then many searches for general philosophical theories of 'how science works' are based on false assumptions about the uniformity of science, and in particular the assumption that there are all these 'theories' in the history of science about which we can say something general and substantive. Perhaps, then, in eliminating *theory*, we can improve the questions philosophers of science are asking quite dramatically.

Finally it is important to emphasize how theory eliminativism departs from other examples of eliminativism in the literature. There have been many calls for eliminativism over the years, with targets including *cause*, *species*, *race*, *concept*, *innate*, *consciousness*, *emotion*, *belief* (and other such concepts in 'folk' psychology), *hard*, *science*, *truth*, and more recently *explanation*. Usually the argument is that the concept in question doesn't correspond to a 'real thing' in the world. Or at least, that there is a plurality of 'things' (or kinds of thing) that people are using a single term to refer to, which is causing serious disruption to debates. For example, Machery (2009) argues that *concept* is not a natural kind, and many critics of 'folk psychology' have argued that 'beliefs', 'emotions', etc. do not exist. This leads to claims that terms such as 'concept' and 'belief' should be *completely* eliminated from philosophical/scientific discourse. This radical proposal has led to some vehement reactions from critics of eliminativism. Fodor (1987: xii) claims that it would be an 'intellectual catastrophe' to eliminate such time-honoured concepts as *belief*.

Theory eliminativism, however, is crucially different from these other proposals in that it is *pragmatic* and *selective*. I have not argued in this work that 'theories don't exist' or that 'theories are not natural kinds'. I have suggested, here and there, that individual theories are 'fuzzy' in that they do not have definitive or 'canonical' content, and also that the examples of 'theories' covered in the case studies do not have enough in common for us to expect to be able to say things as substantive and general as is usually assumed.[13] These more metaphysical, ontological claims about theories have not been central to my account, although I do think that the case studies provide some warrant for them. But the primary motivations for theory eliminativism are pragmatic: (i) the widespread disagreements about what theories are (and little hope of resolution in the near future), and (ii) the engagement with science and the history of science which results from reformulating one's point such that *theory* is eliminated. This leaves open the possibility that in some debates, where disagreements about *theory* do not affect the debate, or where there aren't any disagreements about theory, and/or where engagement with science and the history of science is either not necessary or already sufficiently good, theory eliminativism should not be applied. It still wouldn't do any *harm* to eliminate *theory*, but it's not clear that it would do any good either, in such circumstances. Critics of eliminativism in general might then be more open to this. It is a substantially weaker claim than many eliminativist proposals: sometimes (quite often), for some debates (a significant number), the benefits of reformulating the debate such that *theory* is eliminated will significantly outweigh the costs.[14] Many of the other calls for eliminativism could perhaps benefit from this sort of weakening. For example, Machery (2009) does argue at one point that it's not enough that *concept* is not a natural kind: some terms are extremely useful even though they don't correspond to a natural kind. Part of the motivation for 'concept eliminativism', we are told, is that it will be so *pragmatically* helpful to eliminate the term 'concept' (e.g. Machery 2009: 238f.). This invites the possibility that concept eliminativism can follow the theory eliminativism suggested here and be selective in the sense that it needn't *always* apply. It will depend on the details of the debate at hand. This weakens the claim being made, but accordingly makes it more defensible, more realistic.

A full examination of eliminativism as a methodological tool, and theory eliminativism in particular, will have to wait for another day.[15] Nevertheless

---

[13] There are interesting parallels between these suggestions regarding 'theories' and other claims which have been made in the context of other eliminativist proposals. See, e.g., Ratcliffe (2008), who argues that all of the things we call 'beliefs' actually refer to a wide range of significantly different psychological states. Using the single word 'belief' makes things look much simpler than they really are. Similarly, I suggest, with 'theory'.

[14] In addition, one might wish to restrict (initially, at least) the application of theory eliminativism to debates about the properties of and relations between particular, named theories, as opposed to debates about theories in general.

[15] See Vickers (forthcoming) for some further considerations.

I hope to have done enough to indicate that it can be a powerful strategy and an important alternative to conceptual analysis. It deserves more consideration and application in philosophy. Especially since eliminativists do not need to make overly strong claims that the referent of the term in question 'doesn't exist', or that the corresponding concept should *never* be employed.

8.4 Concluding Thoughts

I have looked at eight cases of 'inconsistent science' in the history of science, many of which are widely cited in the literature as examples of internally inconsistent scientific theories. I have argued that, if one looks more carefully at the relevant history of science, most of these cases are not really inconsistent in any significant sense after all. And to reconstruct such cases *as* inconsistent sets of propositions is a highly dubious move. Often when this is done the motivation seems to be to find an application for a paraconsistent logic, and *not* to say something interesting or important about how real science works, or even could work. Authors glibly list these cases as 'widely known examples of inconsistent scientific theories'. It tells us something important about how philosophy of science works—and how philosophy of science can fail—if something so misleading can become a completely accepted 'fact' in the community.

One explanation of this is that still today, fifty years after Kuhn's *Structure*, too few philosophers of science are getting their hands sufficiently dirty with science and the history of science. This is even true of cases like Aristotle's theory of motion and the early calculus, where the inconsistencies are so obvious that one ought to be extremely sceptical of any claim that there was ever any serious scientific commitment to the relevant inconsistent propositions. There are just too many philosophers failing to read up on the relevant history, partly because there is an unfortunate culture of publishing philosophy of science which makes no attempt to engage with the real scientific or historical facts. Instead it is enough for a philosopher to call his or her account a 'rational reconstruction', as if this in itself is enough to justify any disconnect, however glaring, between the claims being made and the real history of science.

Theory eliminativism helps to expose some of the mistakes which have been made, and also re-shapes debates such that the mistakes are at least partly rectified. Not only does it bring philosophers to engage more seriously with science and history of science, but in doing that it also reveals a diversity in science which often isn't appreciated. That is, many of the case studies investigated here reveal science working in quite radically different ways (even though the cases are nearly all from physics!), and inconsistencies which do exist take on a variety of different characteristics, dependent upon various factors. Calling them all 'inconsistent theories' acts to cover up these important differences, and encourages misguided projects in philosophy of science. This doesn't mean that there are *no* interesting similarities between the cases: in §8.2 I discussed some of these similarities, and the corresponding generalizations which can be made about inconsistency in science. But the more far reaching conclusions concern lessons about the relationship between philosophy of science, history of science, and science itself.

What is the next stop for theory eliminativism? There is nothing special about debates about inconsistency which should persuade us that it will only be effective there. It is reasonable to assume that in any debate which puts weight on the concept *scientific theory* theory eliminativism can be beneficial. And most debates in philosophy of science put considerable weight on this concept. Thus it is possible that with further investigation theory eliminativism could be applied quite widely. Indeed, a quite startling transformation of philosophy of science could ensue.

Bibliography

Abraham, M. (1904): 'Die Grundhypothesen der Elektronentheorie', *Physikalische Zeitschrift* **5**: 576–79.

Alexander, H. G. (1956): *The Leibniz-Clarke Correspondence*. Manchester: Manchester University Press.

Andrews, C. L. (1947): 'Diffraction Pattern of a Circular Aperture at Short Distances', *Physical Review* **71** (11): 777–86.

Arnold, V. I. (1977): *Mathematical Methods of Classical Mechanics*. Berlin: Springer.

Arthur, R. T. W. (1995): 'Newton's Fluxions and Equably Flowing Time', *Studies in History and Philosophy of Science* **26** (2): 323–51.

—— (2008): 'Leery Bedfellows: Newton and Leibniz on the Status of Infinitesimals', in U. Goldenbaum and D. Jesseph (eds), *Infinitesimal Differences: Controversies between Leibniz and his Contemporaries*. Berlin: Walter de Gruyter, 7–30.

—— (2009): 'Actual Infinitesimals in Leibniz's Early Thought', in M. Kulstad, M. Laerke, and D. Snyder (eds), *The Philosophy of the Young Leibniz*. Stuttgart: Franz Steiner Verlag, 11–28.

Baggini, J. and Stangroom, J. (2006): *Do You Think What You Think You Think?* London: Granta Books.

Bartelborth, T. (1989): 'Is Bohr's Model of the Atom Inconsistent?', in P. Weingartner and G. Schurz (eds), *Philosophy of the Natural Sciences, Proceedings of the 13th International Wittgenstein Symposium*. Vienna: Hölder Pichler Tempsky, 220–3.

Bassler, O. B. (2008): 'An Enticing (Im)Possibility: Infinitesimals, Differentials, and the Leibnizian Calculus', in U. Goldenbaum and D. Jesseph (eds), *Infinitesimal Differences: Controversies between Leibniz and his Contemporaries*. Berlin: Walter de Gruyter, 135–51.

Batterman, R. (1995): 'Theories Between Theories: Asymptotic Limiting Intertheoretic Relations', *Synthese* **103**: 171–201.

Baumann, J. J. (1869): *Die Lehre von Raum, Zeit und Mathematik in der neueren Philosophie*, Vol. II. Berlin: Georg Reimer.

Belot, G. (2007): 'Is Classical Electrodynamics an Inconsistent Theory?', *Canadian Journal of Philosophy* **37**: 263–82.

Bergia, S. and Navarro, L. (2000): 'On the early history of Einstein's quantization rule of 1917', *Archives Internationales d'Histoire des Sciences* **50**: 321–73.

Berkeley, G. (1734): *The Analyst*, online edition <http://www.maths.tcd.ie/pub/HistMath/People/Berkeley/Analyst/>. This edition edited by David R. Wilkins. Last accessed: 23 February 2013.

Bernoulli, J. (1924 [1691/2]): *Die Differentialrechnung/von Johann Bernoulli. nach der in der Basler Universitätsbibliothek befindlichen Handschrift/übersetzt, mit einem Vorwort und Anmerkungen versehen von Paul Schafheitlin*. Leipzig: Akademische Verlagsgesellschaft.

Bohr, N. (1913a): 'On the Constitution of Atoms and Molecules', *Philosophical Magazine* **26** (6): 1–25, 476–502, and 857–75.

—— (1913b): 'The Spectra of Helium and Hydrogen', *Nature* **92**: 231–2.

—— (1918): 'On the Quantum Theory of Line Spectra', reproduced in *N. Bohr: Collected Works*, Volume 3. Amsterdam: North Holland Publishing Company, 67–166.

—— (1921): 'Atomic Structure', *Nature*, 24 March.

Bokulich, A. (2006): 'Heisenberg Meets Kuhn: Closed Theories and Paradigms', *Philosophy of Science* **73**: 90–107.

Born, M. and Wolf, E. (1999): *Principles of Optics*, 7th (expanded) edn. Cambridge: Cambridge University Press.

Bos, H. (1974): 'Differentials, Higher-order Differentials and the Derivative in the Leibnizian Calculus', *Archive for History of Exact Sciences* **14**: 1–90.

—— (1980): 'Newton, Leibniz and the Leibnizian Tradition', in I. Grattan-Guiness (ed.), *From the Calculus to Set Theory, 1630–1910*. London: Gerald Duckworth & Co., 49–93.

Boyer, C. B. (1949): *The History of the Calculus and its Conceptual Development*, New York: Dover.

—— (1968): *A History of Mathematics*. New York/Chichester: John Wiley & Sons.

Braithwaite, R. (1953): *Scientific Explanation*. Cambridge: Cambridge University Press.

Brandom, R. (1994): *Making it Explicit*. Cambridge, MA: Harvard University Press.

Breger, H. (2008): 'Leibniz's Calculation with Compendia', in U. Goldenbaum and D. Jesseph (eds), *Infinitesimal Differences: Controversies between Leibniz and his Contemporaries*. Berlin: Walter de Gruyter, 185–98.

Brigandt, I. (2010): 'Scientific Reasoning is Material Inference: Combining Confirmation, Discovery, and Explanation', *International Studies in the Philosophy of Science* **24** (1): 31–43.

Brook, R. J. (1973): *Berkeley's Philosophy of Science*. The Hague: Martinus Nojhoff.

Brown, B. (1990): 'How to be Realistic about Inconsistency in Science', *Studies in History and Philosophy of Science* **21**: 281–94.

Brown, B. (1992): 'Old Quantum Theory: A Paraconsistent Approach', *PSA 1992* **2**: 397–411.
—— (2002): 'Approximate Truth: A Paraconsistent Account', in J. Meheus (ed.), *Inconsistency in Science*. Dordrecht: Kluwer, 81–103.
Brown, B. and Priest, G. (2004): 'Chunk and Permeate, A Paraconsistent Inference Strategy. Part I: The Infinitesimal Calculus', *Journal of Philosophical Logic* **33** (4): 379–88.
Brown, J. (2000): 'Thought Experiments', in W. H. Newton-Smith (ed.), *A Companion to the Philosophy of Science*. Oxford: Blackwell, 528–31.
Brzeziński, J. and Nowak, L. (eds) (1992): *Idealization III: Approximation and Truth*. Amsterdam: Rodopi.
Bueno, O. (2006): 'Why Inconsistency is not Hell: Making Room for Inconsistency in Science', in Erik J. Olsson (ed.), *Knowledge and Inquiry: Essays on the pragmatism of Isaac Levi*. Cambridge: Cambridge University Press, 70–86.
Burian, R. M. (1977): 'More Than a Marriage of Convenience: On the Inextricability of History and Philosophy of Science', *Philosophy of Science* **44**: 1–42.
Cajori, F. (1917): 'Discussion of Fluxions: From Berkeley to Woodhouse', *The American Mathematical Monthly* **24** (4): 145–54.
Carnap, R. (1939): *Foundations of Logic and Mathematics*. Chicago: University of Chicago Press.
—— (1950): *Logical Foundations of Probability*. London: Routledge and Kegan Paul.
—— (1967): *The Logical Structure of the World: Pseudo-Problems in Philosophy*, 2nd edn. London: Routledge and Kegan Paul.
Cartwright, N. (1983): *How the Laws of Physics Lie*. Oxford: Clarendon Press.
—— Fleck, L., Cat, J., and Uebel, T. (1996): *Otto Neurath: Philosophy between Science and Politics*. Cambridge: Cambridge University Press.
Cat, J. (2001): 'On Understanding: Maxwell on the Methods of Illustration and Scientific Metaphor', *Studies in History and Philosophy of Modern Physics* **32** (3) 395–441.
Cauchy, A. L. (1821): *Cours d'Analyse de L'Ecole Royale Polytechnique*. Première Partie. Analyse algébrique. Paris: Imprimérie Royale.
Chakravartty, A. (2001): 'The Semantic or Model-Theoretic View of Theories and Scientific Realism', *Synthese* **127**: 325–45.
Clark, M. (2002): *Paradoxes from A to Z*. London and New York: Routledge.
Cleland, C. E. (2001): 'Recipes, Algorithms, and Programs', *Minds and Machines* **11**: 219–37.
—— (2002): 'On Effective Procedures', *Minds and Machines* **12**: 159–79.
Colyvan, M. (2008): 'The Ontological Commitments of Inconsistent Theories', *Philosophical Studies* **141**: 115–23.

—— (2009): 'Applying Inconsistent Mathematics' in O. Bueno and Ø. Linnebo (eds), *New Waves in Philosophy of Mathematics*. Basingstoke: Palgrave Macmillan, 160–72.

Craver, C. (2002): 'Structures of Scientific Theories', in P. Machamer and M. Silberstein (eds), *The Blackwell Guide to the Philosophy of Science*. Malden, MA: Blackwell, 55–79.

Da Costa, N. C. A. and French, S. (1990): 'Belief, Contradiction and the Logic of Self-Deception', *American Philosophical Quarterly* **27** (3): 179–97.

—— and —— (2003): *Science and Partial Truth*, Oxford: Oxford University Press.

Dales, H. and Oliveri, G. (eds) (1998): *Truth in Mathematics*. New York: Clarendon Press.

Darden, L. (1991): *Theory Change in Science: Strategies from Mendelian Genetics*. Oxford: Oxford University Press.

—— and Maull, N. (1977): 'Interfield Theories', *Philosophy of Science* **44** (1): 43–64.

Darrigol, O. (1992): *From c-Numbers to q-Numbers*. Oxford: University of California Press.

—— (2008): 'The Modular Structure of Physical Theories', *Synthese* **162**: 195–223.

Davey, K. (2003): 'Is Mathematical Rigor Necessary in Physics?', *British Journal for the Philosophy of Science* **54**: 439–63.

Dirac, P. A. M. (1938): 'Classical Theory of Radiating Electrons', *Proceedings of the Royal Society A* **167**: 148–69.

Duffin, W. J. (1990): *Electricity and Magnetism*, 4th edn. Maidenhead: McGraw-Hill.

Earman, J. (1986): *A Primer on Determinism*. Dordrecht: Reidel.

—— and Friedman, M. (1973): 'The Meaning and Status of Newton's Law of Inertia and the Nature of Gravitational Forces', *Philosophy of Science* **40** (3): 329–59.

Edwards, C. H. (1979): *The Historical Development of the Calculus*. New York: Springer-Verlag.

Ehrenfest, P. (1917): 'Adiabatic Invariants and the Theory of Quanta', *Philosophical Magazine* **33**: 500–13.

Einstein, A. (1917): *Relativity: the Special and the General Theory*, 15th edn (trans. R. W. Lawson). London: Methuen (1954).

Eisberg, R. and Resnick, R. (1985): *Quantum Physics of Atoms, Molecules, Solids, and Particles*, 2nd edn. New York: Wiley.

Emerson, W. (1743): *The Doctrine of Fluxions: not only Explaining the Elements thereof, but also its Application and Use in the Several Parts of Mathematics and Natural Philosophy*. London: J. Bettenham (*sold by* W. Innys, 2nd edn. corrected and enlarged 1757; 3rd edn. 1768; 4th edn. 1773).

Feyerabend, P. (1978): 'In Defence of Aristotle', in G. Radnitsky and G. Anderson (eds), *Progress and Rationality in Science*. Dordrecht: Reidel.

Feynman, R., Leighton, R., and Sands, M. (1964): *The Feynman Lectures on Physics*, vol. II. Reading, MA: Addison-Wesley.

Finkelstein, D. (1966): 'Matter, Space and Logic', in C. A. Hooker (ed.), *The Logico-algebraic Approach to Quantum Mechanics II. Boston Studies in the Philosophy of Science, Proceedings of the Boston Colloquium for the Philosophy of Science V*. Dordrecht: Kluwer, 199–215.

Fodor, J. (1987): *Psychosemantics: The Problem of Meaning in the Philosophy of Mind*. Cambridge, MA: MIT Press.

Fowler, A. (1913): 'The Spectra of Helium and Hyrogen', *Nature* **92**: 95–6.

Fraser, D. (2009): 'Quantum Field Theory: Underdetermination, Inconsistency, and Idealization', *Philosophy of Science* **76** (4): 536–67.

French, S. (2003): 'A Model-Theoretic Account of Representation', *Philosophy of Science* **70**: 1472–83.

—— (2008): 'The Structure of Theories', in S. Psillos and M. Curd (eds), *The Routledge Companion to the Philosophy of Science*. London/New York: Routledge, 269–80.

—— and Saatsi, J. (2006): 'Realism about Structure: The Semantic View and Nonlinguistic Representations', *Philosophy of Science* **73**: 548–59.

—— and Vickers, P. (2011): 'Are There No Things That Are Scientific Theories?', *British Journal for the Philosophy of Science* **62** (4): 771–804.

Friedrich, B. and Herschbach, D. (2003): 'Stern and Gerlach: How a Bad Cigar Helped Reorient Atomic Physics', *Physics Today*, December: 53–9.

Frisch, M. (2004): 'Inconsistency in Classical Electrodynamics', *Philosophy of Science* **71**: 525–49.

—— (2005a): *Inconsistency, Asymmetry, and Non-Locality*. Oxford: Oxford University Press.

—— (2005b): 'Mechanisms, Principles, and Lorentz's Cautious Realism', *Studies in History and Philosophy of Modern Physics* **36**: 659–79.

—— (2008): 'Conceptual Problems in Classical Electrodynamics', *Philosophy of Science* **75**: 93–105.

Gendler, T. S. (1998): 'Galileo and the Indispensability of Scientific Thought Experiments', *British Journal for the Philosophy of Science* **49**: 397–424.

Giere, R. (1988): *Explaining Science*. London: University of Chicago Press.

Goldberg, S. (1970): 'The Abraham Theory of the Electron: The Symbiosis of Experiment and Theory', *Archive for History of Exact Sciences* **7**: 7–25.

Goldenbaum, U. and Jesseph, D. (eds) (2008): *Infinitesimal Differences: Controversies between Leibniz and his Contemporaries*. Berlin: Walter de Gruyter.

Gould, S. J. (2002): *The Structure of Evolutionary Theory*. London: Belknap Press of Harvard University Press.

Grabiner, J. (1983): 'The Changing Concept of Change: The Derivative from Fermat to Weierstrass', *Mathematics Magazine* **56** (4): 195–206.

—— (1997): 'Was Newton's Calculus a Dead End? The Continental Influence of Maclaurin's Treatise of Fluxions', *The American Mathematical Monthly* **104** (5): 393–410.

Grattan-Guinness, I. (1970): *The Development of the Foundations of Mathematical Analysis from Euler to Riemann*. Cambridge, MA: MIT Press.

—— (1990): *Convolutions in French Mathematics, 1800–1840*. Basel: Birkhäuser Verlag.

Griffiths, D. (1999): *Introduction to Electrodynamics*. Upper Saddle River, NJ: Prentice-Hall.

Guicciardini, N. (1989): *The Development of Newtonian Calculus in Britain, 1700–1800*. Cambridge: Cambridge University Press.

—— (2004): 'Isaac Newton and the Publication of his Mathematical Manuscripts', *Studies in History and Philosophy of Science* **35**: 455–70.

Hanson, N. R. (1962): 'The Irrelevance of History of science to Philosophy of Science', *The Journal of Philosophy* **59**: 574–86.

Hardy, G. H. (1949): *Divergent Series*. Oxford: Clarendon.

Harman, G. (1986): *Change in View*. Cambridge, MA: MIT Press.

Heaviside, O. (1899): *Electromagnetic Theory 2*. London: 'The Electrician' Print. and Publ. Co.

Heilbron, J. and Kuhn, T. (1969): 'The Genesis of the Bohr Atom', *Hitorical Studies in the Physical Sciences* **1**: 211–90.

Henderson, L., Goodman, N. D., Tenenbaum, J. B., and Woodward, J. F. (2010): 'The Structure and Dynamics of Scientific Theories: A Hierarchical Bayesian Perspective', *Philosophy of Science* **77**: 172–200.

Hendry, R. F. (1993): 'Realism, History and the Quantum Theory: Philosophical and Historical Arguments for Realism as a Methodological Thesis'. PhD thesis, LSE.

Hendry, R. and Psillos, S. (2007): 'How to Do Things with Theories: An Interactive View of Language and Models in Science', in J. Brzeziński, A. Klawiter, T. A. F. Kuipers, K. Łastowski, K. Paprzycka, and P. Przybysz (eds), *The Courage of Doing Philosophy: Essays Dedicated to Leszek Nowak*. Amsterdam/New York: Rodopi, 59–115.

Hettema, H. (1995): 'Bohr's Theory of the Atom 1913–1923: A Case Study in the Progress of Scientific Research Programmes', *Studies in History and Philosophy of Modern Physics* **26**: 307–23.

Heurtley, J. C. (1973): 'Scalar Rayleigh-Sommerfeld and Kirchhoff Diffraction Integrals: A Comparison of Exact Evaluations for Axial Points', *Journal of the Optical Society of America* **63** (8): 1003–8.

Horvath, M. (1986): 'On the Attempts made by Leibniz to Justify his Calculus', *Studia Leibnitiana* **18** (1): 60–71.

Horwich, P. (1991): 'On the Nature and Norms of Theoretical Commitment', *Philosophy of Science* **58**: 1–14.

Howard, D. (2011): 'Philosophy of Science and the History of Science', in S. French and J. Saatsi (eds), *The Continuum Companion to the Philosophy of Science*. London/New York: Continuum Press, 55–71.

Hunter, G. (1971): *Metalogic*. Chicago: University of Chicago Press.

Hutchison, K. (1993): 'Is Classical Mechanics Really Time-Reversible and Deterministic?', *British Journal for the Philosophy of Science* **44** (2): 307–23.

Jackson, J. (1962): *Classical Electrodynamics*, 1st edn. New York: John Wiley and Sons.

——(1999): *Classical Electrodynamics*, 3rd edn. New York: John Wiley and Sons.

Jaki, S. L. (1969): *The Paradox of Olbers' Paradox*. New York: Herder and Herder.

——(1979): 'Das Gravitations-Paradox des unendlichen Universums'. *Sudhoffs Archiv* **63**: 105–22.

Jammer, M. (1966): *The Conceptual Development of Quantum Mechanics*. London: McGraw-Hill.

Jaśkowski, S. (1948): 'Rachunek zdán dla systemóv dedukcyjnych sprzecznych', *Studia Societatis Scientiarum Torunensis* Sectio A, I: 171–2.

Jeans, J. H. (1924): *Report on Radiation and the Quantum Theory*. London: Fleetway Press.

Jesseph, D. M. (1993): *Berkeley's Philosophy of Mathematics*. Chicago: University of Chicago Press.

——(1998): 'Leibniz on the Foundations of the Calculus: The Question of the Reality of Infinitesimal Magnitudes', *Perspectives on Science* **6** (1/2): 6–40.

——(2008): 'Truth in Fiction: Origins and Consequences of Leibniz's Doctrine of Infinitesimal Magnitudes', in U. Goldenbaum and D. Jesseph (eds), *Infinitesimal Differences: Controversies between Leibniz and his Contemporaries*. Berlin: Walter de Gruyter, 215–33.

Kapitan, T. (1982): 'On the Concept of Material Consequence', *History and Philosophy of Logic* **3**: 193–211.

Katz, M. and Sherry, D. (forthcoming): 'Leibniz's Infinitesimals: Their Fictionality, their Modern Implementations, and their Foes from Berkeley to Russell and Beyond', *Erkenntnis*, Online First. Online ISSN: 1572-8420. DOI: 10.1007/s10670-012-9370-y.

——Schaps, D., and Shnider, S. (forthcoming 2013): 'Almost Equal: The Method of Adequality from Diophantus to Fermat and Beyond', *Perspectives on Science*

21 (3). Online access: <http://arxiv.org/abs/1210.7750>. Last accessed: 23 February 2013.

Kenat, R. (1987): 'Physical Interpretation: Eddington, Idealization and Stellar Structure Theory'. PhD thesis, University of Maryland.

Ketland, J. (2005): 'From a Deflationary Point of View', *Notre Dame Philosophical Reviews*. Review of Paul Horwich, *From a Deflationary Point of View*. Oxford: Oxford University Press.

Kirchhoff, G. (1882): 'On the Theory of Light Rays' (trans. Royal Prussian Academy of Sciences, Berlin) **15** Berlin Mb, 641–69. Published again the following year as: *Annalen der Physik* **2** (18): 663–95.

Kitcher, P. (1973): 'Fluxions, Limits and Infinite Littlenesse: A Study of Newton's Presentation of the Calculus', *Isis* **64** (1): 33–49.

——(1983): *The Nature of Mathematical Knowledge*. Oxford: Oxford University Press.

Klein, M. (1985): *Paul Ehrenfest. The Making of a Theoretical Physicist*. Amsterdam: North-Holland Physics Publishing.

Knobloch, E. (2002): 'Leibniz's Rigorous Foundation of Infinitesimal Geometry by Means of Riemannian Sums', *Synthese* **133** (1/2): 59–73.

Kochiras, H. (2013): 'Causal Language and the Structure of Force in Newton's *System of the World*', *HOPOS: The Journal of the International Society for the History of Philosophy of Science* **3** (2).

Korolev, A. (2007a): 'Indeterminism, Asymptotic Reasoning, and Time Irreversibility in Classical Physics', *Philosophy of Science* **74**: 943–56.

——(2007b): 'The Norton-Type Lipschitz-Indeterministic Systems and Elastic Phenomena: Indeterminism as an Artefact of Infinite Idealizations'. Available on the Philsci Archive: <http://philsci-archive.pitt.edu/4314/>. Last accessed 23 February 2013.

Kragh, H. (1999): *Quantum Generations. A History of Physics in the Twentieth Century*. Princeton: Princeton University Press.

——(2001): 'The Electron, the Protyle, and the Unity of Matter', in J. Buchwald and A. Warwick (eds), *Histories of the Electron: The Birth of Microphysics*. Cambridge, MA: MIT Press, 195–226.

——(2012): *Niels Bohr and the Quantum Atom: The Bohr Model of Atomic Structure 1913–1925*. Oxford: Oxford University Press.

Kramers, H. (1919): 'Intensities of Spectral Lines', in *Collected Scientific Papers*. Amsterdam: North Holland, 1956.

—— and Holst, H. (1923): *The Atom and the Bohr Theory of its Structure*. London: Gyldendal.

Kuhn, T. (1962): *The Structure of Scientific Revolutions*. Chicago: University of Chicago Press.

—— (1977): 'Objectivity, Value Judgement, and Theory Choice', in T. Kuhn, *The Essential Tension*. Chicago: University of Chicago Press.

Kythe, P. K. and Puri, P. (2002): *Computational Methods for Linear Integral Equations*. Dordrecht: Springer.

Lakatos, I. (1966): 'Cauchy and the Continuum: The Significance of Non-Standard Analysis for the History and Philosophy of Mathematics', *Mathematical Intelligencer* **1** (1978): 151–61.

—— (1970a): 'Falsification and the Methodology of Scientific Research Programs', in I. Lakatos and A. Musgrave (eds), *Criticism and the Growth of Knowledge*. Cambridge: Cambridge University Press, 91–195.

—— (1970b): 'History of Science and Its Rational Reconstructions', in R. C. Buck and R. S. Cohen (eds), *PSA 1970*. Dordrecht: D. Reidel Publ. Co., 91–136.

Landau, L. and Lifshitz, E. (1951): *The Classical Theory of Fields* (trans. Morton Hamermesh). Cambridge, MA: Addison-Wesley.

—— and —— (1962): *Course of Theoretical Physics: The Classical Theory of Fields*, Vol. 2, 2nd edn (trans. Morton Hamermesh). London: Pergamon Press.

Lange, M. (2002): *An Introduction to the Philosophy of Physics: Locality, Fields, Energy and Mass*. Oxford: Blackwell.

Laudan, L. (1977): *Progress and its Problems*. Ewing, NJ: University of California Press.

—— Donovan, A., Laudan, R., Barker, P., Brown, H., Leplin, J., Thagard, P., and Wykstra, S. (1986): 'Scientific Change: Philosophical Models and Historical Research', *Synthese* **69**: 141–223.

Laugwitz, D. (1989): 'Definite Values of Infinite Sums', *Archive for History of Exact Sciences* **39**: 195–245.

Lavine, S. (1994): *Understanding the Infinite*. Cambridge, MA: Harvard University Press.

Layzer, D. (1954): 'On the Significance of Newtonian Cosmology', *The Astronomical Journal* **59**: 268–70.

Levey, S. (2008): 'Archimedes, Infinitesimals and the Law of Continuity: On Leibniz's Fictionalism', in U. Goldenbaum and D. Jesseph (eds), *Infinitesimal Differences: Controversies between Leibniz and his Contemporaries*. Berlin: Walter de Gruyter, 107–33.

Lindsay, R. B. (1927): 'Note on "Pendulum" Orbits in Atomic Models', *Proceedings of the National Academy of Sciences of the United States of America* **13**: 413–19.

Lorentz, H. A. (1916): *The Theory of Electrons and Its Applications to the Phenomena of Light and Radiant Heat*, 2nd edn. Leipzig: B. G. Teubner.

Lycan, W. G. (1994): *Modality and Meaning.* Dordrecht: Kluwer.

Mc Cormmach, R. (1970): 'Einstein, Lorentz, and the Electron Theory', *Historical Studies in the Physical Sciences* **2**: 41–87.

McCrea, W. H. (1955): 'On the Significance of Newtonian Cosmology', *The Astronomical Journal* **60**: 271–4.

MacFarlane, J. (2000): 'What Does It Mean to Say that Logic is Formal?' PhD thesis, University of Pittsburgh, PA.

Machery, E. (2009): *Doing Without Concepts.* Oxford: Oxford University Press.

Maclaurin, C. (1742): *A Treatise of Fluxions in Two Books.* Edinburgh: T. Ruddimans.

Maher, P. (1990): 'Acceptance without Belief', in A. Fine, M. Forbes, and L. Wessels (eds), *PSA 1990*, Vol. 1, East Lansing, MI: Philosophy of Science Association, 381–92.

Malament, D. (1995): 'Is Newtonian Cosmology Really Inconsistent?', *Philosophy of Science* **62**: 489–510.

Mancosu, P. (1989): 'The Metaphysics of the Calculus: A Foundational Debate in the Paris Academy of Sciences, 1700–1706', *Historia Mathematica* **16**: 224–48.

—— (1996): *Philosophy of Mathematics and Mathematical Practice in the Seventeenth Century*, Oxford: Oxford University Press.

Marchand, E. W. and Wolf, E. (1966): 'Consistent Formulation of Kirchhoff's Diffraction Theory', *Journal of the Optical Society of America* **56** (12): 1712–22.

Marchildon, L. (2002): *Quantum Mechanics: From Basic Principles to Numerical Methods and Applications.* Berlin: Springer-Verlag.

Mártin, H. O. and Tsallis, C. (1981): 'Renormalization Group Specific Heat and Magnetization of the Ising Ferromagnet in Cubic and Hypercubic Lattices', *Zeitschrift für Physik B* **44** (4): 325–31.

Meheus, J. (ed.) (2002): *Inconsistency in Science.* Dordrecht: Kluwer Academic Publishers.

Mehra, J. and Rechenberg, H. (1982): *Historical Development of Quantum Mechanics.* Dordrecht: Springer.

—— and —— (2000a): *The Quantum Theory of Planck, Einstein, Bohr and Sommerfeld: Its Foundation and the Rise of Its Difficulties.* Dordrecht: Springer.

—— and —— (2000b): *The Fundamental Equations of Quantum Mechanics, 1925–1926: The Reception of the New Quantum Mechanics, 1925–1926*, Volume 4. Dordrecht: Springer.

Merleau-Ponty, J. (1977): 'Laplace as a Cosmologist', in W. Yourgrau and A. D. Breck (eds), *Cosmology, History and Theology.* New York: Plenum Press, 283–91.

Merleau-Ponty, J. and Morando, B. (1976): *The Rebirth of Cosmology*. New York: Alfred A. Knopf.

Miller, A. I. (1981): *Albert Einstein's Special Theory of Relativity: Emergence (1905) and Early Interpretation (1905–1911)*. Reading, MA: Addison-Wesley.

Millikan, R. A. (1917): *The Electron: Its Isolation and Measurement and the Determination of some of its Properties*. Chicago: University of Chicago Press.

Morrison, M. (2007): 'Where Have All the Theories Gone?', *Philosophy of Science* **74**: 195–228.

Muller, F. A. (2007): 'Inconsistency in Classical Electrodynamics?', *Philosophy of Science* **74**: 253–77.

Nagel, E. (1961): *The Structure of Science*. London: Routledge and Kegan Paul.

Nagel, F. (2008): 'Nieuwentijt, Leibniz, and Jacob Hermann on Infinitesimals', in U. Goldenbaum and D. Jesseph (eds), *Infinitesimal Differences: Controversies between Leibniz and his Contemporaries*. Berlin: Walter de Gruyter, 199–214.

Newton, I. (1971): *The Mathematical Papers of Isaac Newton*, Volume IV, *1674–1684*. D. T. Whiteside (ed.). Cambridge: Cambridge University Press.

—— (1981): *The Mathematical Papers of Isaac Newton*, Volume VIII, *1697–1722*. D. T. Whiteside (ed.), Cambridge: Cambridge University Press.

—— (1999 [1687]): *The Principia, Mathematical Principles of Natural Philosophy* (trans. I. B. Cohen and A. Whitman, assisted by Julia Budenz). Berkeley, LA/London: University of California Press.

Newton-Smith, W. H. (1981): *The Rationality of Science*. Boston: Routledge and Kegan Paul.

Nickles, T. (1977): 'Heuristics and Justification in Scientific Research: Comments on Shapere', in F. Suppe (ed.), *The Structure of Scientific Theories*. Illinois: University of Illinois Press, 518–65.

—— (2002): 'From Copernicus to Ptolemy: Inconsistency and Method', in J. Meheus (ed.), *Inconsistency in Science*. Dordrecht: Kluwer Academic Publishers, 1–33.

Niiniluoto, I. (1984): *Is Science Progressive?* Dordrecht/Boston: D. Reidel.

North, J. D. (1965): *The Measure of the Universe: A History of Modern Cosmology*. Oxford: Clarendon.

Norton, J. (1987): 'The Logical Inconsistency of the Old Quantum Theory of Black Body Radiation', *Philosophy of Science* **54**: 327–50.

—— (1993): 'A Paradox in Newtonian Cosmology', in D. Hull, M. Forbes, and K. Okruhlik (eds), *PSA 1992*, Vol. 2. East Lansing, MI: Philosophy of Science Association, 412–20.

—— (1995): 'The Force of Newtonian Cosmology: Acceleration is Relative', *Philosophy of Science* **62**: 511–22.

—— (1999): 'The Cosmological Woes of Newtonian Gravitation Theory', in H. Goenner et al. (eds), *The Expanding Worlds of General Relativity*, Einstein Studies, vol. 7, 271–323.

—— (2000): 'How We Know About Electrons', in R. Nola and H. Sankey (eds), *After Popper, Kuhn and Feyerabend*. Dordrecht: Kluwer, 67–97.

—— (2002): 'A Paradox in Newtonian Gravitation Theory II', in J. Meheus (ed.), *Inconsistency in Science*. Dordrecht: Kluwer Academic Publishers, 185–95.

—— (2008): 'The Dome: An Unexpectedly Simple Failure of Determinism', *Philosophy of Science* **75**: 786–98.

—— (2011): 'History of Science and the Material Theory of Induction: Einstein's Quanta, Mercury's Perihelion', *European Journal for Philosophy of Science* **1** (1): 3–27.

Pais, A. (1972): 'The Early History of the Theory of the Electron', in A. Salam and E. P. Wigner (eds), *Aspects of Quantum Theory*. Cambridge: Cambridge University Press, 79–94.

—— (1991): *Niels Bohr's Times*. Oxford: Oxford University Press.

Pauli, W. (1926): 'Quantentheorie', in H. Geiger and K. Scheel (eds), *Handbuch der Physik*. Berlin: Springer, vol. 23, 1–278. (Reprinted in W. Pauli (1964): *Collected Scientific Papers*, Vol. 1, ed. R. Kronig and V. F. Weisskopf. New York: Interscience Publishers, 1964: 271.)

Piccinini, G. and Scott, S. (2006): 'Splitting Concepts', *Philosophy of Science* **73**: 390–409.

Pincock, C. (2010): 'Exploring the Boundaries of Conceptual Evaluation', *Philosophia Mathematica* **18** (III): 106–36.

Poincaré, H. (1892): *Théorie mathématique de la lumière*. Paris: Gauthier Villars.

Popper, K. (1940): 'What is Dialectic?' *Mind* **49**: 403–36.

—— (1959): *The Logic of Scientific Discovery*. London: Hutchinson and Co.

Priest, G. (2002): 'Inconsistency and the Empirical Sciences', in J. Meheus (ed.), *Inconsistency in Science*. Dordrecht: Kluwer, 119–28.

Priest, G. and Routley, R (1983): *On Paraconsistency*, Research Report 13, Logic Group, Research School of Social Sciences, Australian National University.

Psillos, S. and Curd, M. (eds) (2008): *The Routledge Companion to the Philosophy of Science*. London/New York: Routledge.

Ratcliffe, M. (2008): 'Farewell to Folk Psychology: A Response to Hutto', *International Journal of Philosophical Studies* **16** (3): 445–51.

Read, S. (1994): 'Formal and Material Consequence', *Journal of Philosophical Logic* **23**: 247–65.

Reichenbach, H. (1938): *Experience and Prediction: An Analysis of the Foundations and the Structure of Knowledge*. Chicago: University of Chicago Press.

Reichenbach, H. (1951): *The Rise of Scientific Philosophy*. Berkeley: University of California Press.

Rescher, N. (1955): 'Leibniz' Conception of Quantity, Number, and Infinity', *The Philosophical Review* **64** (1): 108–14.

—— (1973): *The Primacy of Practice*. Oxford: Blackwell.

Rohrlich, F. (2007): *Classical Charged Particles*. Singapore: World Scientific.

—— and Hardin, L. (1983): 'Established Theories', *Philosophy of Science* **50**: 603–17.

Saatsi, J. and Vickers, P. (2011): 'Miraculous Success? Inconsistency and Untruth in Kirchhoff's Diffraction Theory', *British Journal for the Philosophy of Science* **62** (1): 29–46.

Salmon, W. (1965): *The Foundations of Scientific Inference*. Pittsburgh: University of Pittsburgh Press.

Sarukkai, S. (2005): 'Revisiting the "Unreasonable Effectiveness" of Mathematics', *Current Science* **88** (3): 415–23.

Scerri, E. R. (1993): 'Correspondence and Reduction in Chemistry', in S. French and H. Kamminga (eds), *Correspondence, Invariance and Heuristics: Essays in Honour of Heinz Post*. Dordrecht: Kluwer.

Schickore, J. (2011): 'More Thoughts on HPS: Another 20 Years Later', *Perspectives on Science* **19** (4): 453–81.

Schott, G. (1918): 'On Bohr's Hypothesis of Stationary States of Motion and the Radiation from an Accelerated Electron', *Philosophical Magazine* **36**: 243–61.

Schrenk, M. (2004): 'Galileo vs. Aristotle on Free Falling Bodies', *Logical Analysis and History of Philosophy*, volume 7: *History and Philosophy of Nature*, 1–11.

Seeliger, H. (1895): 'Über das Newton'sche Gravitationsgesetz', *Astronomische Nachrichten* **137** (3273): 129–36.

Shapere, D. (1969): 'Notes toward a Post-Positivistic Interpretation of Science', in P. Achinstein and S. F. Barker (eds), *The Legacy of Logical Positivism*. Baltimore: MD: Johns Hopkins University Press, 115–60.

—— (1977): 'Scientific Theories and Their Domains', in F. Suppe (ed.), *The Structure of Scientific Theories*. Illinois: University of Illinois Press, 518–65.

—— (1984a): *Reason and the Search for Knowledge*. Dordrecht: Reidel.

—— (1984b): 'Objectivity, Rationality and Scientific Change', *PSA 1984* **2**: 637–63.

Sherry, D. (1987): 'The Wake of Berkeley's Analyst: Rigor Mathematicae?', *Studies in History and Philosophy of Science* **18** (4): 455–80.

Sitte, B. and Egbers, C. (2000): 'Experimental Investigations on Nonlinear Behaviour of Baroclinic Waves', in R. J. Adrian (ed.), *Laser Techniques Applied to Fluid Mechanics: Selected Papers from the 9th International Symposium, Lisbon, Portugal, July 13–16, 1998*. Berlin: Springer, 315–36.

Smith, J. (1988): 'Inconsistency and Scientific Reasoning', *Studies in History and Philosophy of Science* **19**: 429–45.
Smith, P. (2007): *An Introduction to Gödel's Theorems*. Cambridge: Cambridge University Press.
Sommerfeld, A. (1916): 'Zur Quantentheorie der Spektrallinien', *Annalen der Physik* **51**: 1–94, 125–67.
Spohn, H. (2004): *Dynamics of Charged Particles and their Radiation Field*, Cambridge: Cambridge University Press.
Suppe, F. (ed.) (1977): *The Structure of Scientific Theories*, 2nd edn. Illinois: University of Illinois Press.
—— (1989): *The Semantic Conception of Theories and Scientific Realism*. Illinois: University of Illinois Press.
Suppes, P. (1957): *Introduction to Logic*. New York: Van Nostrand.
—— (1967): 'What is a scientific theory?', in S. Morgenbesser (ed.), *Philosophy of Science Today*. New York: Basic Books, 55–67.
Taylor, C. (1977): *Hegel*. Cambridge: Cambridge University Press.
Toulmin, S. (1972): *Human Understanding*. Oxford: Clarendon Press.
Van Fraassen, B. (1980): *The Scientific Image*. Oxford: Oxford University Press.
Van Vleck, J. H. (1926): *Quantum Principles and Line Spectra*. Washington, DC: National Research Council of the National Academy of Sciences.
Vickers, P. (2008): 'Frisch, Muller and Belot on an Inconsistency in Classical Electrodynamics', *British Journal for the Philosophy of Science* **59** (4): 767–92.
—— (2009a): 'Was Newtonian Cosmology Really Inconsistent?', *Studies in History and Philosophy of Modern Physics* **40** (3): 197–208.
—— (2009b): 'Can Partial Structures Accommodate Inconsistent Science?', *Principia* **13** (2): 233–50.
—— (2012): 'Historical Magic in Old Quantum Theory?', *European Journal for Philosophy of Science* **2** (1): 1–19.
—— (forthcoming): 'Scientific Theory Eliminativism', *Erkenntnis*. DOI: 10.1007/s10670-013-9471-2.
Votsis, I. (2011): 'The Prospective Stance in Realism', *Philosophy of Science* **78** (5): 1223–34.
Warmbrod, K. (1999): 'Logical constants', *Mind* **108**: 503–38.
Whiteside, D. T. and Newton, I. (1967): *The Mathematical Papers of Isaac Newton*, Vol. 1, 1664–1666. Cambridge: Cambridge University Press.
Wilson, M. (2006): *Wandering Significance*. Oxford: Oxford University Press.
—— (2009): 'Determinism and the Mystery of the Missing Physics', *British Journal for the Philosophy of Science* **60**: 173–93.

Wilson, M. (2013): 'What is "Classical Mechanics" Anyway?', in R. Batterman (ed.), *The Oxford Handbook of Philosophy of Physics*. Oxford: Oxford University Press.

Yaghjian, A. D. (1992): *Relativistic Dynamics of a Charged Sphere: Updating the Lorentz-Abraham Model*. Berlin: Springer.

Zinkernagel, H. (2010): 'Causal Fundamentalism in Physics', in M. Suárez, M. Dorato, and M. Rédei (eds), *EPSA Philosophical Issues in the Sciences*. Dordrecht: Springer, 311–22.

Index

Abraham, M. 81, 103n.18, 204
Abraham-Lorentz equation 81, 103n.18
adiabatic principle 59, 64–8
algorithm 147, 150–9, 161, 163–5, 169–70, 173–4, 176, 180, 182–3, 185, 187–9, 219, 223, 232, 244
angular momentum 42, 49
anti-realism 144–5, 231
approximations 53–8, 62–3, 68–9, 73–4, 88, 91, 94, 104–7, 117, 147–8, 208, 212–13, 215–16, 232, 241–2, 245, 247
  internalizing the approximation 58, 212–13
Aristotle 2, 8, 192–4, 198, 216, 219, 221–4, 226–8, 252
  Aristotle's theory of motion 192–6, 198, 216, 219, 221–4, 226–8, 252
Arrhenius, S. 137, 139–40

bad questions 13, 15, 34, 38, 71, 74–5, 224–5, 233
Belot, G. 31, 36, 76–7n.1, 80–1n.5, 84, 86–101, 106–8, 199, 203n.2, 248
Bentley, R. 138, 222
Berkeley, G. 146n.1, 147–51, 153–5, 158, 161, 163–4n.18, 169–74, 179, 181, 183, 223
Bernoulli, Johann 147, 167, 175–9, 181, 183–5, 187, 189, 191n.36, 223, 232
Bohr, N. 25–6, 31–2, 36, 39–76, 106, 109, 144–5, 157, 191–3, 198, 207, 212, 219–22, 224–6, 230, 232–3, 238–41, 245, 247–8
Boyer, C. B. 146–8n.2, 152, 158, 170, 176, 179
burden of proof 45, 52, 57

capitalism 229–231n.2
Carnap, R. 8, 20
Cartwright, N. 11, 24–5, 28n.9,
Cauchy, A. 127n.16, 137, 142–3, 154n.8, 157
CED, *see* classical physics
Clarke, S. 228–9

classical logic 7, 9–11, 18n.1, 48, 69n.23, 185, 189, 241
classical physics 43–46, 49, 234
  classical electrodynamics (CED) 26–7, 30–1, 36, 41, 48, 50–5, 59, 61–3, 72–3 75–109, 145, 199–207, 224–6, 228, 231–2, 240, 245, 247
  classical electromagnetism, *see* classical electrodynamics
  classical mechanics 26–7n.8, 43, 242–3, 248–9n.11
  classical versus quantum 42–63, 71–3
  electrostatics 42, 50–1n.7, 200–5, 207
conceptual analysis 17, 19–20, 26, 30–4, 100, 252
concept eliminativism 250–1
  see also eliminativism
concept splitting 108–9n.21
  see also pluralism
conceptual weight 19, 30, 35, 85, 248, 250, 253
content-driven approach 6–10, 238–42
  see also logic-driven approach
convergence and non-convergence of series 116–18, 124, 127nn.16–17, 128–130, 132, 137–43, 161–4n.18
  see also Grandi's series
correspondence principle, *see* old quantum theory
Coulomb field/force 42, 50–2, 78–81, 83, 200–7
  see also self-field of a charged particle; self-force of a charged particle; Maxwell's equations
credence, *see* epistemic confidence

debate reformulation 32n.10, 37–8, 85–6, 217, 227, 243–53
Debye, P. 66–7, 70, 144, 230, 232
deductive inferences, *see* truth-preserving inferences
De Morgan, A. 142
diffraction 55n.9, 208–15, 220, 231, 235, 238–9, 245

Dirac, P. 6, 8, 61n.16, 202, 207–8, 234
domain of phenomena 20, 26, 31, 36, 50, 59, 89–93, 95–7, 99, 101–3, 105–7, 109, 132, 145, 222, 224–6, 235, 239
 see also pointedly grouped propositions
doxastic commitment 4, 24, 29–31, 34, 47, 49–53, 58, 60, 62, 67, 70, 73–4, 91, 99, 103–4, 106–7, 115, 123, 134, 159, 175–9, 182–4, 187, 194, 202, 204–8, 212–14, 217, 230–1, 241, 244–5, 247, 249

early calculus 8, 146–91, 196, 207, 220, 223–4, 231–2, 238
 fluxions 8, 146, 149n.3, 151, 158–63, 170–4, 179, 181n.32, 186–8, 232
 infinitesimals 149n.3, 150, 156, 158–90, 231–2
ECQ, see ex contradictione quodlibet
Ehrenfest, P. 36, 59, 67, 72n.24
Einstein, A. 39–40, 121n.12, 129–30, 200, 204–6
electrons 33, 40–53, 57, 59, 65–6, 68, 70–1, 80–2, 98, 102, 199–208, 216, 220, 223–4, 226–8, 230–2, 234
 electron spin, see spin
 as extended particles 78–81, 98, 199, 202, 206, 231, 234
 discrete energies/orbits 43–6, 49, 226
 internal structure of electrons 79, 81, 199, 202, 204, 206, 208, 234
 Poincaré counter-pressure 205–6, 234
 as point particles 50, 77–9, 82, 98, 106, 199–208, 228, 231, 234
electromagnetic radiation 39, 42–3, 48–53, 64–6, 69, 76–109, 199–208, 220, 224, 226
 and atoms 39, 42–3, 48–53, 64–6
 synchrotron radiation 102–4, 225
electrostatics, see classical physics
elementary particles 82, 106, 200–8, 228
 see also electrons
eliminativism 28–32, 37–8, 42, 44, 73, 75, 84–5, 91, 94–5, 100, 107–9, 112, 133, 150–1, 188, 203, 218, 242–53
 and pluralism 108–9
 concept eliminativism 250–2
energy conservation 77, 81, 83, 91, 93–4, 97, 101, 105
epistemic confidence 3, 40, 52–8, 63, 69, 104n.19, 182, 212–15, 240–1, 245
error bars 56

Euclidean space 112, 114, 120, 124, 134, 198
 see also non-Euclidean space
Euler, L. 141, 175–176
evolution theory, see theory of evolution
ex contradictione quodlibet 7–10, 29, 48, 54–5, 63, 69–70, 73, 104, 106–7, 145, 165, 182, 185, 189, 207, 212, 214–15, 217, 240–1
 see also logical closure
explanation 93, 96, 106–7, 152–3, 157–8, 179–80, 198, 224, 250
external inconsistency, see mutual inconsistency

Feyerabend, P. 8, 69, 146, 190
Feynman, R. 80, 99, 200, 228
fictionalism 165–9, 182, 184–7, 230, 232
French, S. 9, 12, 20–1n.4, 23, 41, 43–4, 49, 146, 190
Frisch, M. 9–12, 22, 28n.9, 30–1, 36, 76–8, 82–108, 199–200, 205–6, 208, 224, 228, 240, 248

Galileo 147, 192–6, 221, 223, 226–9, 247
gauge interpretations 121n.11, 132, 144, 230–3
Goldbach's conjecture 13, 54–5, 69, 74, 216, 241, 245
Gould, S. J. 20, 26–7n.7
Grandi's series 116–18, 128n.19, 141–2
gravitational potential 112–13, 119, 122, 128–31, 135, 137n.25, 142–3

Halley, E. 199
Heaviside, O. 157, 159
Helium 39, 50
Helmholtz, H. von 210
historical relevance 29, 35–6, 42, 45–7, 71, 74, 91, 99, 103, 113, 133, 143, 147, 153, 189, 194, 197, 201, 227
hydrogen 39, 42, 50, 58, 64–71, 144, 219–20
 and the adiabatic principle 64–71, 219
 crossed fields 65, 219
 pendulum orbit 65, 70, 220

idealizations 47, 57, 63, 67–9, 73, 86n.9, 104, 106–7, 209–10, 212, 214, 217, 221, 226, 231, 235, 239–40, 245
 de-idealizing 68, 86n.9, 214, 221, 231, 239–40
implicit premises, see implicit propositions

implicit propositions 33–4, 45–6, 202–4, 233–4
inconsistency
  definition of 17–19, 32–4
  inconsistency resolution 2, 4n.2, 121n.12, 230, 232–3
  versus implausibility 226–30
indeterminacy 113, 118, 121, 123–32, 137–40, 142, 236–7
inference rules 7, 9, 18–19, 63, 185, 190, 236
infinity 117–18, 127n.16, 128–30, 139, 179
infinitesimals 149n.3, 150, 156, 158–61, 163–90, 230–2
  as variables 177–81, 188, 232
integrated history and philosophy of science 25, 35–8, 45, 48, 71, 105, 123, 133, 135, 189–91, 196, 199, 216–17, 241–53
inter-theory relations 5

Jaki, S. 129, 134–6, 139n.27, 196–9

Kant, I. 231n.2
Kaufmann, W. 204
Kelvin, Lord 40, 130
Kepler, J. 199
Kirchhoff, G. 37, 55n.9, 208–17, 220–2, 224, 226, 231, 234–5, 238–41, 245
  theory of diffraction 37, 55n.9, 208–17, 220–2, 231, 235, 238–41, 245
Kitcher, P. 151–3, 159–61, 163–4n.19, 173, 181n.32, 183
Kuhn, T. 3, 20, 38, 224, 242–3, 246, 252

Lakatos, I. 8, 20, 32–3, 38, 41, 49, 59, 63n.18, 146, 151, 190, 243
Laplace, P. S. 122n.13, 135, 143
Leibniz, G. W. 6, 8, 37, 142, 145–9n.3, 151–4, 160n.15, 164–9, 174–91n.36, 228–32
Le Verrier, U. 135
LFE, see Lorentz force equation
l'Hôpital, G. de 175–7, 182
logical closure 12, 48, 108
  see also ex contradictione quodlibet
logical explosion, see ex contradictione quodlibet
logic-driven approach 6–10, 12, 219, 238–42
  see also content-driven approach

Lorentz, H. 76–7, 79, 81–2, 103n.18, 200, 203–6, 234–5
  see also Lorentz force equation
  theory of the electron 203–6, 234–5
Lorentz force equation 50, 76–107, 199

Maclaurin, C. 172–4, 186n.34, 188
Marxism 229
material inferences 19, 33–4
mathematics 2–3, 12–13, 27, 36, 54, 116n.6, 117n.7, 127, 127nn.16–17, 135, 137–9n.28, 142, 144–7, 151, 154, 157, 186, 190–1, 236–8
  inconsistency in 190
  philosophy of 116n.6, 117n.7, 236–8
  role in physics 139n.28, 145
  truth preserving inferences 236–8
Maxwell's equations 26–7, 40, 51, 55n.9, 76, 82–3, 87, 88n.10, 91–4, 97, 101, 103n.18, 105, 107, 206, 241
  Coulomb's law 50–2
  Gauss's law 50–2
mechanics, see classical mechanics
metaphilosophy 37, 73, 106–7, 217, 219, 252–3
Millikan, R. A. 51
Minkowski, H. 206
miscommunications 13, 15, 27, 31, 76, 85, 99, 107, 230, 246
Morrison, M. 5, 20–1, 26–7
Muller, F. 31, 36, 53, 76, 80, 84–6, 88, 106–7, 202, 207–8, 240, 247
mutual inconsistency 4, 4n.2, 40, 49, 63, 229–30

Newton, I. 6, 26, 27n.8, 36, 96, 103, 110, 112–14, 120, 124–6, 127n.17, 129, 131, 134, 136–9n.28, 141, 143, 146–53, 158–69, 171–4, 179, 181n.32, 182–4n.33, 186–7, 189–90, 193, 222, 232, 235–7, 243
Newtonian cosmology 26, 36, 110–45, 154, 157, 191, 196, 198, 219, 222–6, 230–3, 235, 237–9, 241, 245
  modern theory of 111, 121n.11, 132, 233
Newtonian mechanics, see classical mechanics
Newton's law of gravitational attraction 96, 112–14, 118, 124–7, 129, 131–4, 136, 139n.28, 141n.29, 145
  see also Newtonian cosmology

Newton's laws of motion 26, 103, 112–14, 118, 124–5, 127n.17, 132, 134, 136, 139n.28, 145, 243
  see also classical mechanics
non-classical logic 8–9, 133, 185–7, 190, 193, 224, 238, 241, 246, 252
non-convergent series, see convergence and non-convergence of series; Grandi's series
non-Euclidean space 198
  see also Euclidean space
non-logical inferences, see material inferences
non-uniformity of science 37–8, 71, 133, 246, 250
Norton, J. 9, 26, 51n.7, 110–12n.2, 114–19n.9, 121n.12, 122, 124–5n.14, 128n.18, 130–1, 136–41n.29, 144, 233, 239, 241, 247–8

Olbers, H. W. M. 129n.20, 196–9, 216, 220, 223–4, 228
old quantum mechanics, see old quantum theory
old quantum theory 25–6, 31–2, 36, 39–76, 91, 103, 106, 109–10, 123, 134, 144–5, 154, 191, 217, 222–3, 226, 230, 232–3, 235, 240, 245, 247–8
  adiabatic principle 64–71, 219
  correspondence principle 50, 59–63n.18, 72–3
  Sommerfeld's developments 58, 60, 64–8, 70, 232

paraconsistent logic, see non-classical logic
Pauli, W. 64–70, 72, 74, 91, 103, 106, 109–10, 123, 134, 144, 154, 157, 191, 217, 219–22, 225, 231–2, 235, 245
pluralism 108–9
  see also eliminativism
Poincaré, H. 205–6, 212–15, 224, 234–5
  Poincaré counter-pressure 205–6, 234
pointedly grouped propositions 29, 31, 34, 36, 45–7, 49, 52, 68, 71, 74, 90–3, 95–7, 103, 105, 107–9, 112, 136, 194, 197, 212, 216, 218, 223, 225, 227
Poisson, S. D. 112–13, 118–23, 130–1, 135, 137n.25, 141–3
  Poisson's equation 112–13, 118–23, 130–5, 137n.25, 141–3

Popper, K. 6–8, 10
potential, see gravitational potential
Poynting vector 87
pragmatic commitment 153, 206–7, 224–5, 236
predictions 4n.2, 39, 58–61, 63, 66–7, 70, 118, 209, 212, 214, 216–17, 220–1, 224, 238–9, 244–5, 247

quantum field theory 237–8n.5
quantum jumps, see quantum transitions
quantum numbers 50, 59, 61, 61n.15, 64–6, 72
  principal quantum number 50, 59, 61, 61 n.15
quantum transitions 42–3, 46–8, 71, 226
  see also classical versus quantum

rational reconstruction 23–5, 36n.11, 42, 45, 103, 105, 110–11, 165n.20, 186–7, 189–90, 196, 201, 223–4, 238, 241, 243n.6, 249n.10, 253
Reichenbach, H. 8, 20
relativity 4n.2, 9, 33, 79, 82, 86n.9, 98, 144, 199–207, 233
  general relativity 4n.2, 9, 144, 233
  special relativity 33, 79, 82, 86n.9, 98, 199–207
reliability of inferences 52–6, 63, 69, 104n.19, 167, 214–15, 239–42, 245
renormalization 202, 207, 238
representing scientific theories 4, 6, 8, 10–12, 15, 20, 21n.4, 22–5, 188
rhetoric 223, 227, 230
Rolle, M. 155, 175, 178, 181, 185
Russell, B. 133
Rutherford, E. 41, 47

science 2–3, 5, 8, 11, 22–6, 37–8, 134–6, 206 n.4, 242–3
  demarcation 134–5, 206n.4, 226
scientific communities 29–31, 35, 39–40, 51, 53, 56–63, 66–8, 71–4, 97, 99–100, 105, 108, 110, 147, 153, 155, 158, 161, 169–71, 174–81, 183, 185, 188, 198, 204, 214, 216–17, 220–2, 232, 245
scientific explanation, see explanation
Seeliger, H. 36, 110, 130, 139, 143–4, 222–3, 225, 239
self-field of a charged particle 78–81, 83, 86, 91–3, 95–7, 105, 199, 224

self-force of a charged particle 78, 80–4, 86, 92–3, 95–8, 101–8, 199, 224–5, 231
semantic approach to theories 10–12, 19–25
Shapere, D. 8, 20, 36, 49, 60, 73, 73n.25, 146, 190, 200–3, 224–5, 228
significance of an inconsistency 31, 36, 49, 83, 91–3, 95, 99–104, 106, 123, 170, 188, 193, 198, 201, 203, 207, 212, 216–17, 222, 225, 238, 245–7, 252
Sommerfeld, A. 58, 60, 64–8, 70, 210, 232
  theory of the atom 60, 64–70, 232
special relativity, *see* relativity
spectral lines 39, 49–50n.6, 59–65, 72–3
spin 32–3, 68
static field, *see* Coulomb field/force
symmetry 79, 113, 120–3, 138–40
synchrotron radiation, *see* electromagnetic radiation
syntactic approach to theories 8, 10–12, 19–25

tacit knowledge 49, 195n.1
  see also implicit propositions
theories of concepts, *see* conceptual analysis
theories of theories 3, 19–29
  see also semantic approach to theories; syntactic approach to theories; representing scientific theories
theory content 13–14, 23, 25–8, 39–41, 46, 85, 90–1, 95, 100, 105, 108, 112, 150, 243, 248
theory eliminativism, *see* eliminativism
theory identity, *see* theory content
theory-ladenness of history 36n.11
theory names 29–32, 36, 85, 218, 230, 249
theory of evolution 4n.2, 20, 27, 27n.7, 229, 249n.10
thermodynamics 4n.2, 198–9, 229
Thomson, J. J. 206n.4
Thomson, W., *see* Kelvin, Lord
thought experiments 193–196
truth-preserving inferences 7, 18–19, 53–4, 56, 63, 106, 117, 120, 142–5, 185, 215–16, 235–7, 240–1

uniformity of science, *see* non-uniformity of science

Varignon, P. 167, 175, 178–81n.32, 188

Zeeman effect 64–6